# Essential Guide to Semiconductors

# Essential Guide to Semiconductors

Editor: John Holland

NY RESEARCH PRESS
New York

Published by NY Research Press
118-35 Queens Blvd., Suite 400,
Forest Hills, NY 11375, USA
www.nyresearchpress.com

Essential Guide to Semiconductors
Edited by John Holland



International Standard Book Number: 978-1-63238-594-9 (Hardback)



**Cataloging-in-Publication Data**

Essential guide to semiconductors / edited by John Holland.
p. cm.
Includes bibliographical references and index.
ISBN 978-1-63238-594-9
1. Semiconductors. 2. Solid state electronics. I. Holland, John.
TK7871.85 .E87 2018
621.381 52--dc23

Printed in China.

# Contents

# Preface

Materials which have electrical conductivity are known as semiconductors. These materials have different properties like being sensitive to heat and light, passing current in one direction, having variable resistance, can emit light, can be used as thermal energy converters, etc. The most common semiconductor materials are germanium, silicon carbide, organic compounds, gallium arsenide, silicon, etc. This book unfolds the innovative aspects of semiconductors which will be crucial for the holistic understanding of the subject matter. Most of the topics introduced in it cover new techniques and the applications of the semiconductor materials. This textbook is meant for students who are looking for an elaborate reference text on semiconductors.

A foreword of all chapters of the book is provided below:

**Chapter 1 -** Semiconductor devices are devices that use electronic conduction. By introducing electric or magnetic field, it is possible to manipulate the conductivity of these devices. The various semiconductor devices are diode, transistor, integrated circuit and solar cell. This is an introductory chapter which will introduce briefly all the significant aspects of semiconductor devices; **Chapter 2 -** The branch of physics that studies nature and the behavior of matter and energy is known as quantum physics. It helps in expanding the knowledge on individual behaviors of subatomic particles. Some of the fields where quantum physics is applied are electronics, cryptography and quantum computing. This chapter will provide an integrated understanding of quantum physics; **Chapter 3 -** Charge carriers are those particles found in electrical conductors that are free to move. Semi conductors require a certain correlation between valence band and the conduction band in order for a charge carrier to carry electric current. In order to completely understand energy bands and energy carriers it is necessary to understand the processes related to it. The following chapter elucidates the varied processes and mechanisms associated with this area of study; **Chapter 4 -** Semiconductors can be of two types, p type and n type. The boundary found inside a single crystal of a semiconductor which binds them together is known as a p-n junction. This chapter discusses P-N junction in a critical manner providing key analysis to the subject matter.

At the end, I would like to thank all the people associated with this book devoting their precious time and providing their valuable contributions to this book. I would also like to express my gratitude to my fellow colleagues who encouraged me throughout the process.

**Editor**

1

# Fundamentals of Semiconductor Devices

Semiconductor devices are devices that use electronic conduction. By introducing electric or magnetic field, it is possible to manipulate the conductivity of these devices. The various semiconductor devices are diode, transistor, integrated circuit and solar cell. This is an introductory chapter which will introduce briefly all the significant aspects of semiconductor devices.

## Semiconductor Device

Semiconductor devices are electronic components that exploit the electronic properties of semiconductor materials, principally silicon, germanium, and gallium arsenide, as well as organic semiconductors. Semiconductor devices have replaced thermionic devices (vacuum tubes) in most applications. They use electronic conduction in the solid state as opposed to the gaseous state or thermionic emission in a high vacuum.

Semiconductor devices are manufactured both as single discrete devices and as *integrated circuits* (ICs), which consist of a number—from a few (as low as two) to billions—of devices manufactured and interconnected on a single semiconductor substrate, or wafer.

Semiconductor materials are useful because their behavior can be easily manipulated by the addition of impurities, known as doping. Semiconductor conductivity can be controlled by the introduction of an electric or magnetic field, by exposure to light or heat, or by the mechanical deformation of a doped monocrystalline grid; thus, semiconductors can make excellent sensors. Current conduction in a semiconductor occurs via mobile or "free" *electrons* and *holes*, collectively known as *charge carriers*. Doping a semiconductor such as silicon with a small amount of impurity atoms, such as phosphorus or boron, greatly increases the number of free electrons or holes within the semiconductor. When a doped semiconductor contains excess holes it is called "p-type", and when it contains excess free electrons it is known as "n-type", where *p* (positive for holes) or *n* (negative for electrons) is the sign of the charge of the majority mobile charge carriers. The semiconductor material used in devices is doped under highly controlled conditions in a fabrication facility, or *fab*, to control precisely the location and concentration of p- and n-type dopants. The junctions which form where n-type and p-type semiconductors join together are called p–n junctions.

### Diode

A semiconductor diode is a device typically made from a single p–n junction. At the junction of a p-type and an n-type semiconductor there forms a depletion region where current conduction is inhibited by the lack of mobile charge carriers. When the device is *forward biased* (connected with the p-side at higher electric potential than the n-side), this depletion region is diminished, allowing for significant conduction, while only very small current can be achieved when the diode is *reverse biased* and thus the depletion region expanded.

Exposing a semiconductor to light can generate electron–hole pairs, which increases the number of free carriers and thereby the conductivity. Diodes optimized to take advantage of this phenomenon are known as *photodiodes*. Compound semiconductor diodes can also be used to generate light, as in light-emitting diodes and laser diodes.

## Transistor

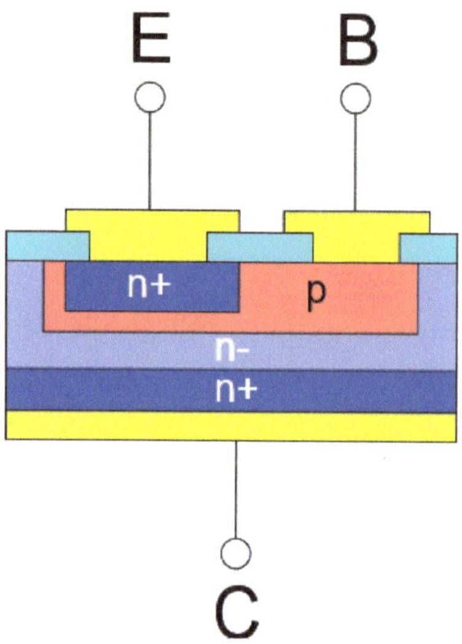


An n–p–n bipolar junction transistor structure

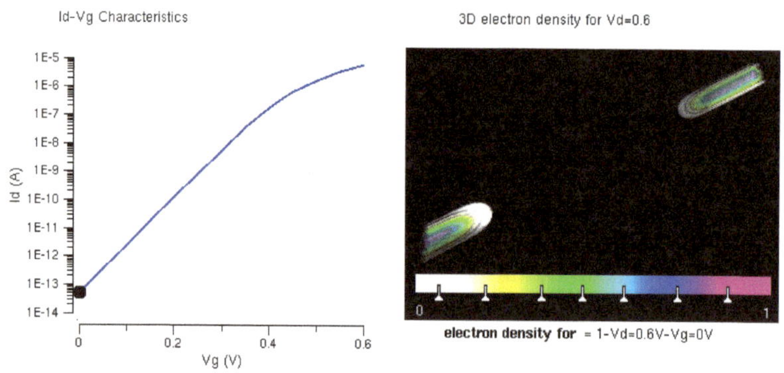


Operation of a MOSFET and its Id-Vg curve. At first, when no gate voltage is applied. There is no inversion electron in the channel, the device is OFF. As gate voltage increase, inversion electron density in the channel increase, current increase, the device turns on

## Bipolar Junction Transistor

Bipolar junction transistors are formed from two p–n junctions, in either n–p–n or p–n–p configuration. The middle, or *base*, region between the junctions is typically very narrow. The other regions, and their associated terminals, are known as the *emitter* and the *collector*. A small current injected through the junction between the base and the emitter changes the properties of the base-collector junction so that it can conduct current even though it is reverse biased. This creates a much larger current between the collector and emitter, controlled by the base-emitter current.

## Field-effect Transistor

Another type of transistor, the field-effect transistor, operates on the principle that semiconductor conductivity can be increased or decreased by the presence of an electric field. An electric field can increase the number of free electrons and holes in a semiconductor, thereby changing its conductivity. The field may be applied by a reverse-biased p–n junction, forming a *junction field-effect transistor* (JFET) or by an electrode insulated from the bulk material by an oxide layer, forming a *metal–oxide–semiconductor field-effect transistor* (MOSFET).

The MOSFET, a solid-state device, is the most used semiconductor device today. The *gate* electrode is charged to produce an electric field that controls the conductivity of a "channel" between two terminals, called the *source* and *drain*. Depending on the type of carrier in the channel, the device may be an *n-channel* (for electrons) or a *p-channel* (for holes) MOSFET. Although the MOSFET is named in part for its "metal" gate, in modern devices polysilicon is typically used instead.

## Semiconductor Device Materials

By far, silicon (Si) is the most widely used material in semiconductor devices. Its combination of low raw material cost, relatively simple processing, and a useful temperature range makes it currently the best compromise among the various competing materials. Silicon used in semiconductor device manufacturing is currently fabricated into boules that are large enough in diameter to allow the production of 300 mm (12 in.) wafers.

Germanium (Ge) was a widely used early semiconductor material but its thermal sensitivity makes it less useful than silicon. Today, germanium is often alloyed with silicon for use in very-high-speed SiGe devices; IBM is a major producer of such devices.

Gallium arsenide (GaAs) is also widely used in high-speed devices but so far, it has been difficult to form large-diameter boules of this material, limiting the wafer diameter to sizes significantly smaller than silicon wafers thus making mass production of GaAs devices significantly more expensive than silicon.

Other less common materials are also in use or under investigation.

Silicon carbide (SiC) has found some application as the raw material for blue light-emitting diodes (LEDs) and is being investigated for use in semiconductor devices that could withstand very high operating temperatures and environments with the presence of significant levels of ionizing radiation. IMPATT diodes have also been fabricated from SiC.

Various indium compounds (indium arsenide, indium antimonide, and indium phosphide) are also being used in LEDs and solid state laser diodes. Selenium sulfide is being studied in the manufacture of photovoltaic solar cells.

The most common use for organic semiconductors is Organic light-emitting diodes.

## List of Common Semiconductor Devices

*Two-terminal devices:*

- DIAC

- Diode (rectifier diode)
- Gunn diode
- IMPATT diode
- Laser diode
- Light-emitting diode (LED)
- Photocell
- Phototransistor
- PIN diode
- Schottky diode
- Solar cell
- Transient-voltage-suppression diode
- Tunnel diode
- VCSEL
- Zener diode

*Three-terminal devices:*

- Bipolar transistor
- Darlington transistor
- Field-effect transistor
- Insulated-gate bipolar transistor (IGBT)
- Silicon-controlled rectifier
- Thyristor
- TRIAC
- Unijunction transistor

*Four-terminal devices:*

- Hall effect sensor (magnetic field sensor)
- Photocoupler (Optocoupler)

## Semiconductor Device Applications

All transistor types can be used as the building blocks of logic gates, which are fundamental in the design of digital circuits. In digital circuits like microprocessors, transistors act as on-off switches; in the MOSFET, for instance, the voltage applied to the gate determines whether the switch is on or off.

Transistors used for analog circuits do not act as on-off switches; rather, they respond to a continuous range of inputs with a continuous range of outputs. Common analog circuits include amplifiers and oscillators.

Circuits that interface or translate between digital circuits and analog circuits are known as mixed-signal circuits.

Power semiconductor devices are discrete devices or integrated circuits intended for high current or high voltage applications. Power integrated circuits combine IC technology with power semiconductor technology, these are sometimes referred to as "smart" power devices. Several companies specialize in manufacturing power semiconductors.

## Component Identifiers

The type designators of semiconductor devices are often manufacturer specific. Nevertheless, there have been attempts at creating standards for type codes, and a subset of devices follow those. For discrete devices, for example, there are three standards: JEDEC JESD370B in United States, Pro Electron in Europe and Japanese Industrial Standards (JIS) in Japan.

## History of Semiconductor Device Development

### Cat's-whisker Detector

Semiconductors had been used in the electronics field for some time before the invention of the transistor. Around the turn of the 20th century they were quite common as detectors in radios, used in a device called a "cat's whisker" developed by Jagadish Chandra Bose and others. These detectors were somewhat troublesome, however, requiring the operator to move a small tungsten filament (the whisker) around the surface of a galena (lead sulfide) or carborundum (silicon carbide) crystal until it suddenly started working. Then, over a period of a few hours or days, the cat's whisker would slowly stop working and the process would have to be repeated. At the time their operation was completely mysterious. After the introduction of the more reliable and amplified vacuum tube based radios, the cat's whisker systems quickly disappeared. The "cat's whisker" is a primitive example of a special type of diode still popular today, called a Schottky diode.

### Metal Rectifier

Another early type of semiconductor device is the metal rectifier in which the semiconductor is copper oxide or selenium. Westinghouse Electric (1886) was a major manufacturer of these rectifiers.

### World War II

During World War II, radar research quickly pushed radar receivers to operate at ever higher frequencies and the traditional tube based radio receivers no longer worked well. The introduction of the cavity magnetron from Britain to the United States in 1940 during the Tizard Mission resulted in a pressing need for a practical high-frequency amplifier.

On a whim, Russell Ohl of Bell Laboratories decided to try a cat's whisker. By this point they had not been in use for a number of years, and no one at the labs had one. After hunting one down at a used radio store in Manhattan, he found that it worked much better than tube-based systems.

Ohl investigated why the cat's whisker functioned so well. He spent most of 1939 trying to grow more pure versions of the crystals. He soon found that with higher quality crystals their finicky behaviour went away, but so did their ability to operate as a radio detector. One day he found one of his purest crystals nevertheless worked well, and interestingly, it had a clearly visible crack near the middle. However as he moved about the room trying to test it, the detector would mysteriously work, and then stop again. After some study he found that the behaviour was controlled by the light in the room–more light caused more conductance in the crystal. He invited several other people to see this crystal, and Walter Brattain immediately realized there was some sort of junction at the crack.

Further research cleared up the remaining mystery. The crystal had cracked because either side contained very slightly different amounts of the impurities Ohl could not remove–about 0.2%. One side of the crystal had impurities that added extra electrons (the carriers of electric current) and made it a "conductor". The other had impurities that wanted to bind to these electrons, making it (what he called) an "insulator". Because the two parts of the crystal were in contact with each other, the electrons could be pushed out of the conductive side which had extra electrons (soon to be known as the *emitter*) and replaced by new ones being provided (from a battery, for instance) where they would flow into the insulating portion and be collected by the whisker filament (named the *collector*). However, when the voltage was reversed the electrons being pushed into the collector would quickly fill up the "holes" (the electron-needy impurities), and conduction would stop almost instantly. This junction of the two crystals (or parts of one crystal) created a solid-state diode, and the concept soon became known as semiconduction. The mechanism of action when the diode is off has to do with the separation of charge carriers around the junction. This is called a "depletion region".

## Development of the Diode

Armed with the knowledge of how these new diodes worked, a vigorous effort began to learn how to build them on demand. Teams at Purdue University, Bell Labs, MIT, and the University of Chicago all joined forces to build better crystals. Within a year germanium production had been perfected to the point where military-grade diodes were being used in most radar sets.

## Development of the Transistor

After the war, William Shockley decided to attempt the building of a triode-like semiconductor device. He secured funding and lab space, and went to work on the problem with Brattain and John Bardeen.

The key to the development of the transistor was the further understanding of the process of the electron mobility in a semiconductor. It was realized that if there were some way to control the flow of the electrons from the emitter to the collector of this newly discovered diode, an amplifier could be built. For instance, if contacts are placed on both sides of a single type of crystal, current will not flow between them through the crystal. However if a third contact could then "inject" electrons or holes into the material, current would flow.

Actually doing this appeared to be very difficult. If the crystal were of any reasonable size, the number of electrons (or holes) required to be injected would have to be very large, making it less than useful as an amplifier because it would require a large injection current to start with. That said, the whole idea of the crystal diode was that the crystal itself could provide the electrons over a very small distance, the depletion region. The key appeared to be to place the input and output contacts very close together on the surface of the crystal on either side of this region.

Brattain started working on building such a device, and tantalizing hints of amplification continued to appear as the team worked on the problem. Sometimes the system would work but then stop working unexpectedly. In one instance a non-working system started working when placed in water. Ohl and Brattain eventually developed a new branch of quantum mechanics, which became known as surface physics, to account for the behaviour. The electrons in any one piece of the crystal would migrate about due to nearby charges. Electrons in the emitters, or the "holes" in the collectors, would cluster at the surface of the crystal where they could find their opposite charge "floating around" in the air (or water). Yet they could be pushed away from the surface with the application of a small amount of charge from any other location on the crystal. Instead of needing a large supply of injected electrons, a very small number in the right place on the crystal would accomplish the same thing.

Their understanding solved the problem of needing a very small control area to some degree. Instead of needing two separate semiconductors connected by a common, but tiny, region, a single larger surface would serve. The electron-emitting and collecting leads would both be placed very close together on the top, with the control lead placed on the base of the crystal. When current flowed through this "base" lead, the electrons or holes would be pushed out, across the block of semiconductor, and collect on the far surface. As long as the emitter and collector were very close together, this should allow enough electrons or holes between them to allow conduction to start.

## The First Transistor

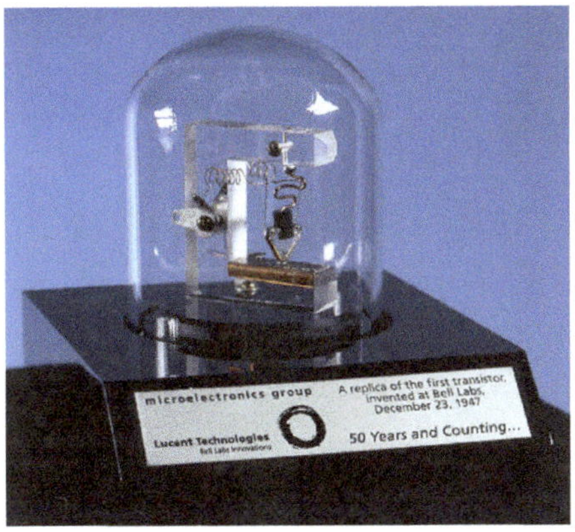


A stylized replica of the first transistor

The Bell team made many attempts to build such a system with various tools, but generally failed. Setups where the contacts were close enough were invariably as fragile as the original cat's whisker detectors had been, and would work briefly, if at all. Eventually they had a practical breakthrough. A piece of gold foil was glued to the edge of a plastic wedge, and then the foil was sliced with a razor at the tip of the triangle. The result was two very closely spaced contacts of gold. When the wedge was pushed down onto the surface of a crystal and voltage applied to the other side (on the base of the crystal), current started to flow from one contact to the other as the base voltage pushed the electrons away from the base towards the other side near the contacts. The point-contact transistor had been invented.

While the device was constructed a week earlier, Brattain's notes describe the first demonstration to higher-ups at Bell Labs on the afternoon of 23 December 1947, often given as the birthdate of the transistor. what is now known as the "p–n–p point-contact germanium transistor" operated as a speech amplifier with a power gain of 18 in that trial. John Bardeen, Walter Houser Brattain, and William Bradford Shockley were awarded the 1956 Nobel Prize in physics for their work.

## Origin of the Term "transistor"

Bell Telephone Laboratories needed a generic name for their new invention: "Semiconductor Triode", "Solid Triode", "Surface States Triode" [*sic*], "Crystal Triode" and "Iotatron" were all considered, but "transistor", coined by John R. Pierce, won an internal ballot. The rationale for the name is described in the following extract from the company's Technical Memoranda (May 28, 1948) calling for votes:

Transistor. This is an abbreviated combination of the words "transconductance" or "transfer", and "varistor". The device logically belongs in the varistor family, and has the transconductance or transfer impedance of a device having gain, so that this combination is descriptive.

## Improvements in Transistor Design

Shockley was upset about the device being credited to Brattain and Bardeen, who he felt had built it "behind his back" to take the glory. Matters became worse when Bell Labs lawyers found that some of Shockley's own writings on the transistor were close enough to those of an earlier 1925 patent by Julius Edgar Lilienfeld that they thought it best that his name be left off the patent application.

Shockley was incensed, and decided to demonstrate who was the real brains of the operation. A few months later he invented an entirely new, considerably more robust, type of transistor with a layer or 'sandwich' structure. This structure went on to be used for the vast majority of all transistors into the 1960s, and evolved into the bipolar junction transistor.

With the fragility problems solved, a remaining problem was purity. Making germanium of the required purity was proving to be a serious problem, and limited the yield of transistors that actually worked from a given batch of material. Germanium's sensitivity to temperature also limited its usefulness. Scientists theorized that silicon would be easier to fabricate, but few investigated this possibility. Gordon K. Teal was the first to develop a working silicon transistor, and his company, the nascent Texas Instruments, profited from its technological edge. From the late 1960s most transistors were silicon-based. Within a few years transistor-based products, most notably easily portable radios, were appearing on the market.

A major improvement in manufacturing yield came when a chemist advised the companies fabricating semiconductors to use distilled rather than tap water: calcium ions present in tap water were the cause of the poor yields. "Zone melting", a technique using a band of molten material moving through the crystal, further increased crystal purity.

The current-voltage characteristics is of prime concern in the study of semiconductor devices with light entering as a third variable in optoelectronics devices.The external characteristics of the device is determined by the interplay of the following internal variables:

1. Electron and hole currents
2. Potential
3. Electron and hole density
4. Doping
5. Temperature

## Semiconductor Equations

The semiconductor equations relating these variables are given below:

Carrier density:

$$n = n_i \exp(\frac{E_{FN} - E_i}{kT}) \quad (1)$$

$$p = n_i \exp(\frac{E_i - E_{FP}}{kT}) \quad (2)$$

where $E_{FN}$ is the electron quasi Fermi level and $E_{FP}$ is the hole quasi Fermi level. These two equations lead to

$$np = n_i^2 \exp(\frac{E_{FN} - E_{FP}}{kT}) \quad (3)$$

In equilibrium $E_{FN} = E_{FP}$ = Constant

Current:

There are two components of current; electron current density $J_N$ and hole current density $J_P$. There are several mechanisms of current flow:

(i) Drift

(ii) Diffusion

(iii) thermionic emission

(iv) tunneling

The last two mechanisms are important often only at the interface of two different materials such as a metal-semiconductor junction or a semiconductor-semiconductor junction where the two semiconductors are of different materials. Tunneling is also important in the case of PN junctions where both sides are heavily doped.

In the bulk of semiconductor , the dominant conduction mechanisms involve drift and diffusion. The current densities due to these two mechanisms can be written as

$$J_N = qn\mu_N\varepsilon + qD_N\frac{dn}{dx} \quad (4)$$

$$J_P = qn\mu_P\varepsilon + qD_P\frac{dP}{dx} \quad (5)$$

where $\mu_N, \mu_P$ are electron and hole mobilities respectively and $D_N, D_P$ are their diffusion constants.

Potential:

The potential and electric field within a semiconductor can be defined in the following ways:

(*i*) $\Psi = -\frac{E_c}{q} + \text{constant}; \varepsilon = \frac{1}{q}\frac{dE_c}{dX}$

(*ii*) $\Psi = -\frac{E_V}{q} + \text{constant}; \varepsilon = \frac{1}{q}\frac{dE_V}{dX}$

(*iii*) $\Psi = -\frac{E_i}{q} + \text{constant}; \varepsilon = \frac{1}{q}\frac{dE_i}{dX}$

(*iv*) $\Psi = -\frac{E_o}{q} + \text{constant}; \varepsilon = \frac{1}{q}\frac{dE_o}{dX}$

All these definitions are equivalent and one or the other may be chosen on the basis of convenience. The potential is related to the carrier densities by the Poisson equation: -

$$\frac{\partial^2\psi}{\partial x^2} = -\frac{q}{\varepsilon}(p - n + N_D^+ - N_A^-) \quad (6)$$

where the last two terms represent the ionized donor and acceptor density.

## Continuity Equations

These equations are basically particle conservation equations:

$$\textit{Electron continuity equation}: \frac{\partial n}{\partial t} = \frac{1}{q}\frac{\partial J_N}{\partial x} + G_N - R_N \quad (7)$$

$$\textit{Hole continuity equation}: \frac{\partial p}{\partial t} = -\frac{1}{q}\frac{\partial J_P}{\partial x} + G_P - R_P \quad (8)$$

Where G and R represent carrier generation and recombination rates.Equations (1-8) will form the basis of most of the device analysis that shall be discussed later on. These equations require models for mobility and recombination along with models of contacts and boundaries.

## Analysis Flow

Like most subjects, the analysis of semiconductor devices is also carried out by starting from simpler problems and gradually progressing to more complex ones as described below:

(i) Analysis under zero excitation i.e. equilibrium.

(ii) Analysis under constant excitation: in other words dc or static characteristics.

(iii) Analysis under time varying excitation but with quasi-static approximation dynamic characteristics.

(iv) Analysis under time varying excitation: non quasi-static dynamic characteristics.

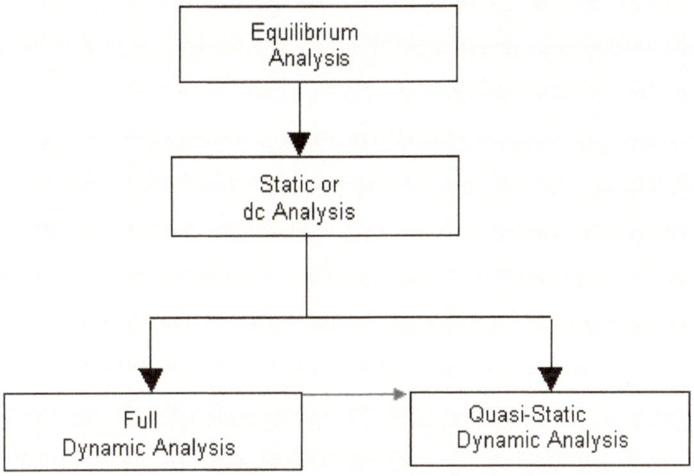


Even though there is zero external current and voltage in equilibrium, the situation inside the device is not so trivial. In general, voltages, charges and drift-diffusion current components at any given point within the semiconductor may not be zero.

Equilibrium in semiconductors implies the following:

$(i)\, steady\, state: \dfrac{\partial Z}{\partial t} = 0$

Where Z is any physical quantity such as charge, voltage electric field etc

(ii) no net electrical current and thermal currents:

Since current can be carried by both electrons and holes, equilibrium implies zero values for both net electron current and net hole current. The drift and diffusion components of electron and hole currents need not be zero.

(iii) Constant Fermi energy: $\dfrac{dE_F}{dX} = 0$

The only equations that are relevant (others being zero!) for analysis in equilibrium are:

$$\frac{\partial^2 \psi_0}{\partial x^2} = -\frac{q}{\varepsilon}(p_o - n_o + N_D^+ - N_A^-)$$

Poisson Eq:

$$n_o = n_i \exp(\frac{E_F - E_i}{kT})$$

$$p_o = n_i \exp(\frac{E_i - E_F}{kT})$$

$$n_o p_o = n_i^2$$

In equilibrium, there is only one independent variable out of the three variables: $n_o, p_o, \psi_o$

If one of them is known, all the rest can be computed from the equations listed above. We shall take this independent variable to be potential.

The analysis problem in equilibrium is therefore determination of potential or equivalently, energy band diagram of the semiconductor device.

This is the reason why we begin discussions of all semiconductor devices with a sketch of its energy band diagram in equilibrium.

## Energy Band Diagram

This diagram in qualitative form is sketched by following the following procedure:

1. The semiconductor device is imagined to be formed by bringing together the various distinct semiconductor layers, metals or insulators of which it is composed. The starting point is therefore the energy band diagram of all the constituent layers.
2. The band diagram of the composite device is sketched using the fact that after equilibrium, the Fermi energy is the same everywhere in the system. The equalization of the Fermi energy is accompanied with transfer of electrons from regions of higher Fermi energy to region of lower Fermi energy and viceversa for holes.
3. The redistribution of charges results in electric field and creation of potential barriers in the system. These effects however are confined only close to the interface between the layers. The regions which are far from the interface remain as they were before the equilibrium

Analysis in equilibrium: Solution of Poisson's Equation with appropriate boundary Conditions -

Non-equilibrium analysis:

- The electron and hole densities are no longer related together by the inverse relationship of Eq. (5) but through complex relationships involving all three variables Y , , p
- The three variables are in general independent of each other in the sense that a knowledge of two of them does not lead automatically to a knowledge of the third.
- The concept of Fermi energy is no longer valid but new quantities called the quasi-Fermi levels are used and these are not in general constant.

- For static or dc analysis, the continuity equation becomes time independent so that only ordinary differential equations need to be solved.
- For dynamic analysis however, the partial differential equations have to be solved increasing the complexity of the analysis.

## Analysis of Semiconductor Devices

There are two complementary ways of studying semiconductor devices:

(i) Through numerical simulation of the semiconductor equations.

(ii) Through analytical solution of semiconductor equations.

- There are a variety of techniques used for device simulation with some of them starting from the drift diffusion formalism outlined earlier, while others take a more fundamental approach starting from the Boltzmann transport equation instead.
- In general, the numerical approach gives highly accurate results but requires heavy computational effort also.
- The output of device simulation in the form of numerical values for all internal variables requires relatively larger effort to understand and extract important relationships among the device characteristics.

## Analytical Techniques

These techniques aim at arriving at closed form symbolic expressions linking important device characteristics. They have the following uses:

(i) They provide basic understanding of device operation

(ii) The devices models obtained are useful for use in circuit analysis

- Analytical solution of semiconductor equations under both equilibrium and non-equilibrium conditions is often a difficult task and is carried out using judicious use of assumptions to obtain simplification.
- For example, assumptions that electron or hole population in a given region is in quasi-equilibrium is frequently used to simplify the analysis.
- At other times, the quasi-static approximation is used to reduce the problem of solving partial differential equation in dynamic analysis to solution of ordinary differential equations only.
- Since, each assumption limits the range of validity of the obtained results, the goal in analytical modeling is to use as few assumptions as are necessary to obtain results which are general, accurate and at the same time simple enough to comprehend and use for circuit analysis.
- It can rightfully be said that the art of analytical modeling is the art of making assumptions!

- Since the requirements of generality, accuracy and simplicity are often - mutually conflicting in nature, one or the other has to be sacrificed.
- The first to be sacrificed is often generality. Instead of obtaining a single model applicable over the entire range of excitation, a set of models for different regions or modes of operation is derived. In each region of operation, a different set of assumptions is used to obtain a simple result.
- As expected, this sectioning of device operation into regions creates problems at the boundaries by causing discontinuity in device characteristics.

## Semiconductor

A semiconductor material has an electrical conductivity value falling between that of a conductor, such as copper, and an insulator, such as glass. Their resistance decreases as their temperature increases, which is behavior opposite to that of a metal. Their conducting properties may be altered in useful ways by the deliberate, controlled introduction of impurities ("doping") into the crystal structure, which lowers its resistance but also permits the creation of semiconductor junctions between differently-doped regions of the extrinsic semiconductor crystal. The behavior of charge carriers which include electrons, ions and electron holes at these junctions is the basis of diodes, transistors and all modern electronics.

Semiconductor devices can display a range of useful properties such as passing current more easily in one direction than the other, showing variable resistance, and sensitivity to light or heat. Because the electrical properties of a semiconductor material can be modified by doping, or by the application of electrical fields or light, devices made from semiconductors can be used for amplification, switching, and energy conversion.

The modern understanding of the properties of a semiconductor relies on quantum physics to explain the movement of charge carriers in a crystal lattice. Doping greatly increases the number of charge carriers within the crystal. When a doped semiconductor contains mostly free holes it is called "p-type", and when it contains mostly free electrons it is known as "n-type". The semiconductor materials used in electronic devices are doped under precise conditions to control the concentration and regions of p- and n-type dopants. A single semiconductor crystal can have many p- and n-type regions; the p–n junctions between these regions are responsible for the useful electronic behavior.

Although some pure elements and many compounds display semiconductor properties, silicon, germanium, and compounds of gallium are the most widely used in electronic devices. Elements near the so-called "metalloid staircase", where the metalloids are located on the periodic table, are usually used as semiconductors.

Some of the properties of semiconductor materials were observed throughout the mid 19th and first decades of the 20th century. The first practical application of semiconductors in electronics was the 1904 development of the Cat's-whisker detector, a primitive semiconductor diode widely used in early radio receivers. Developments in quantum physics in turn allowed the development of the transistor in 1947 and the integrated circuit in 1958.

## Properties

### Variable conductivity

Semiconductors in their natural state are poor conductors because a current requires the flow of electrons, and semiconductors have their valence bands filled, preventing the entry flow of new electrons. There are several developed techniques that allow semiconducting materials to behave like conducting materials, such as doping or gating. These modifications have two outcomes: n-type and p-type. These refer to the excess or shortage of electrons, respectively. An unbalanced number of electrons would cause a current to flow through the material.

### Heterojunctions

Heterojunctions occur when two differently doped semiconducting materials are joined together. For example, a configuration could consist of p-doped and n-doped germanium. This results in an exchange of electrons and holes between the differently doped semiconducting materials. The n-doped germanium would have an excess of electrons, and the p-doped germanium would have an excess of holes. The transfer occurs until equilibrium is reached by a process called recombination, which causes the migrating electrons from the n-type to come in contact with the migrating holes from the p-type. A product of this process is charged ions, which result in an electric field.

### Excited electrons

A difference in electric potential on a semiconducting material would cause it to leave thermal equilibrium and create a non-equilibrium situation. This introduces electrons and holes to the system, which interact via a process called ambipolar diffusion. Whenever thermal equilibrium is disturbed in a semiconducting material, the amount of holes and electrons changes. Such disruptions can occur as a result of a temperature difference or photons, which can enter the system and create electrons and holes. The process that creates and annihilates electrons and holes are called generation and recombination.

### Light emission

In certain semiconductors, excited electrons can relax by emitting light instead of producing heat. These semiconductors are used in the construction of light-emitting diodes and fluorescent quantum dots.

### Thermal energy conversion

Semiconductors have large thermoelectric power factors making them useful in thermoelectric generators, as well as high thermoelectric figures of merit making them useful in thermoelectric coolers.

## Materials

Silicon crystals are the most common semiconducting materials used in microelectronics and photovoltaics

A large number of elements and compounds have semiconducting properties, including:

- Certain pure elements are found in Group 14 of the periodic table; the most commercially important of these elements are silicon and germanium. Silicon and germanium are used here effectively because they have 4 valence electrons in their outermost shell which gives them the ability to gain or lose electrons equally at the same time.
- Binary compounds, particularly between elements in Groups 13 and 15, such as gallium arsenide, Groups 12 and 16, groups 14 and 16, and between different group 14 elements, e.g. silicon carbide.
- Certain ternary compounds, oxides and alloys.
- Organic semiconductors, made of organic compounds.

Most common semiconducting materials are crystalline solids, but amorphous and liquid semiconductors are also known. These include hydrogenated amorphous silicon and mixtures of arsenic, selenium and tellurium in a variety of proportions. These compounds share with better known semiconductors the properties of intermediate conductivity and a rapid variation of conductivity with temperature, as well as occasional negative resistance. Such disordered materials lack the rigid crystalline structure of conventional semiconductors such as silicon. They are generally used in thin film structures, which do not require material of higher electronic quality, being relatively insensitive to impurities and radiation damage.

## Preparation of Semiconductor Materials

Almost all of today's electronic technology involves the use of semiconductors, with the most important aspect being the integrated circuit (IC), which are found in laptops, scanners, cell-phones, etc. Semiconductors for ICs are mass-produced. To create an ideal semiconducting material, chemical purity is paramount. Any small imperfection can have a drastic effect on how the semiconducting material behaves due to the scale at which the materials are used.

A high degree of crystalline perfection is also required, since faults in crystal structure (such as dislocations, twins, and stacking faults) interfere with the semiconducting properties of the material. Crystalline faults are a major cause of defective semiconductor devices. The larger the crystal, the more difficult it is to achieve the necessary perfection. Current mass production processes use crystal ingots between 100 and 300 mm (4 and 12 in) in diameter which are grown as cylinders and sliced into wafers.

There is a combination of processes that is used to prepare semiconducting materials for ICs. One process is called thermal oxidation, which forms silicon dioxide on the surface of the silicon. This is used as a gate insulator and field oxide. Other processes are called photomasks and photolithography. This process is what creates the patterns on the circuity in the integrated circuit. Ultraviolet light is used along with a photoresist layer to create a chemical change that generates the patterns for the circuit.

Etching is the next process that is required. The part of the silicon that was not covered by the photoresist layer from the previous step can now be etched. The main process typically used today is called plasma etching. Plasma etching usually involves an etch gas pumped in a low-pressure chamber to create plasma. A common etch gas is chlorofluorocarbon, or more commonly known Freon. A high radio-frequency voltage between the cathode and anode is what creates the plasma in the chamber. The silicon wafer is located on the cathode, which causes it to be hit by the positively charged ions that are released from the plasma. The end result is silicon that is etched anisotropically.

The last process is called diffusion. This is the process that gives the semiconducting material its desired semiconducting properties. It is also known as doping. The process introduces an impure atom to the system, which creates the p-n junction. In order to get the impure atoms embedded in the silicon wafer, the wafer is first put in a 1100 degree Celsius chamber. The atoms are injected in and eventually diffuse with the silicon. After the process is completed and the silicon has reached room temperature, the doping process is done and the semiconducting material is ready to be used in an integrated circuit.

## Physics of Semiconductors

### Energy Bands and Electrical Conduction

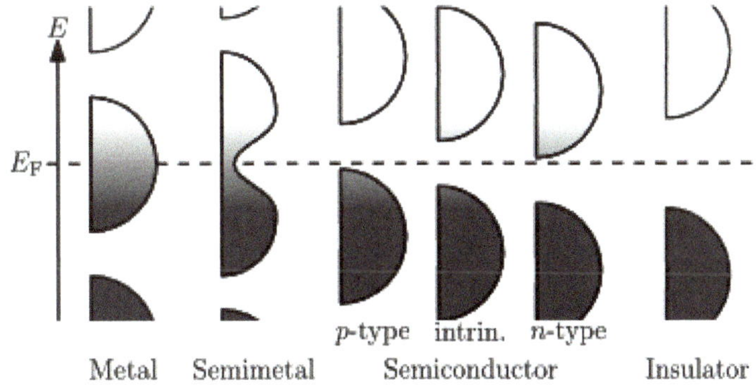


Filling of the electronic states in various types of materials at equilibrium. Here, height is energy while width is the density of available states for a certain energy in the material listed. The shade follows the Fermi–Dirac distribution (black = all states filled, white = no state filled). In metals and semimetals the Fermi level $E_F$ lies inside at least one band. In insulators and semiconductors the Fermi level is inside a band gap; however, in semiconductors the bands are near enough to the Fermi level to be thermally populated with electrons or holes.

Semiconductors are defined by their unique electric conductive behavior, somewhere between that of a conductor and an insulator. The differences between these materials can be understood in terms of the quantum states for electrons, each of which may contain zero or one electron (by the Pauli exclusion principle). These states are associated with the electronic band structure of the material. Electrical conductivity arises due to the presence of electrons in states that are delocalized (extending through the material), however in order to transport electrons a state must be *partially filled*, containing an electron only part of the time. If the state is always occupied with an electron, then it is inert, blocking the passage of other electrons via that state. The energies of these quantum states are critical, since a state is partially filled only if its energy is near the Fermi level.

High conductivity in a material comes from it having many partially filled states and much state delocalization. Metals are good electrical conductors and have many partially filled states with energies near their Fermi level. Insulators, by contrast, have few partially filled states, their Fermi levels sit within band gaps with few energy states to occupy. Importantly, an insulator can be made to conduct by increasing its temperature: heating provides energy to promote some electrons across the band gap, inducing partially filled states in both the band of states beneath the band gap (valence band) and the band of states above the band gap (conduction band). An (intrinsic) semiconductor has a band gap that is smaller than that of an insulator and at room temperature significant numbers of electrons can be excited to cross the band gap.

A pure semiconductor, however, is not very useful, as it is neither a very good insulator nor a very good conductor. However, one important feature of semiconductors (and some insulators, known as *semi-insulators*) is that their conductivity can be increased and controlled by doping with impurities and gating with electric fields. Doping and gating move either the conduction or valence band much closer to the Fermi level, and greatly increase the number of partially filled states.

Some wider-band gap semiconductor materials are sometimes referred to as semi-insulators. When undoped, these have electrical conductivity nearer to that of electrical insulators, however they can be doped (making them as useful as semiconductors). Semi-insulators find niche applications in micro-electronics, such as substrates for HEMT. An example of a common semi-insulator is gallium arsenide. Some materials, such as titanium dioxide, can even be used as insulating materials for some applications, while being treated as wide-gap semiconductors for other applications.

## Charge carriers (Electrons and Holes)

The partial filling of the states at the bottom of the conduction band can be understood as adding electrons to that band. The electrons do not stay indefinitely (due to the natural thermal recombination) but they can move around for some time. The actual concentration of electrons is typically very dilute, and so (unlike in metals) it is possible to think of the electrons in the conduction band of a semiconductor as a sort of classical ideal gas, where the electrons fly around freely without being subject to the Pauli exclusion principle. In most semiconductors the conduction bands have a parabolic dispersion relation, and so these electrons respond to forces (electric field, magnetic field, etc.) much like they would in a vacuum, though with a different effective mass. Because the electrons behave like an ideal gas, one may also think about conduction in very simplistic terms such as the Drude model, and introduce concepts such as electron mobility.

For partial filling at the top of the valence band, it is helpful to introduce the concept of an electron hole. Although the electrons in the valence band are always moving around, a completely full valence band is inert, not conducting any current. If an electron is taken out of the valence band, then the trajectory that the electron would normally have taken is now missing its charge. For the purposes of electric current, this combination of the full valence band, minus the electron, can be converted into a picture of a completely empty band containing a positively charged particle that moves in the same way as the electron. Combined with the *negative* effective mass of the electrons at the top of the valence band, we arrive at a picture of a positively charged particle that responds to electric and magnetic fields just as a normal positively charged particle would do in vacuum, again with some positive effective mass. This particle is called a hole, and the collection of holes in the valence band can again be understood in simple classical terms (as with the electrons in the conduction band).

## Carrier Generation and Recombination

When ionizing radiation strikes a semiconductor, it may excite an electron out of its energy level and consequently leave a hole. This process is known as *electron–hole pair generation*. Electron-hole pairs are constantly generated from thermal energy as well, in the absence of any external energy source.

Electron-hole pairs are also apt to recombine. Conservation of energy demands that these recombination events, in which an electron loses an amount of energy larger than the band gap, be accompanied by the emission of thermal energy (in the form of phonons) or radiation (in the form of photons).

In some states, the generation and recombination of electron–hole pairs are in equipoise. The number of electron-hole pairs in the steady state at a given temperature is determined by quantum statistical mechanics. The precise quantum mechanical mechanisms of generation and recombination are governed by conservation of energy and conservation of momentum.

As the probability that electrons and holes meet together is proportional to the product of their amounts, the product is in steady state nearly constant at a given temperature, providing that there is no significant electric field (which might "flush" carriers of both types, or move them from neighbour regions containing more of them to meet together) or externally driven pair generation. The product is a function of the temperature, as the probability of getting enough thermal energy to produce a pair increases with temperature, being approximately $\exp(-E_G/kT)$, where $k$ is Boltzmann's constant, $T$ is absolute temperature and $E_G$ is band gap.

The probability of meeting is increased by carrier traps—impurities or dislocations which can trap an electron or hole and hold it until a pair is completed. Such carrier traps are sometimes purposely added to reduce the time needed to reach the steady state.

## Doping

The conductivity of semiconductors may easily be modified by introducing impurities into their crystal lattice. The process of adding controlled impurities to a semiconductor is known as *doping*. The amount of impurity, or dopant, added to an *intrinsic* (pure) semiconductor varies its level of conductivity. Doped semiconductors are referred to as *extrinsic*. By adding impurity to the pure semiconductors, the electrical conductivity may be varied by factors of thousands or millions.

A 1 $cm^3$ specimen of a metal or semiconductor has of the order of $10^{22}$ atoms. In a metal, every atom donates at least one free electron for conduction, thus 1 $cm^3$ of metal contains on the order of $10^{22}$ free electrons, whereas a 1 $cm^3$ sample of pure germanium at 20 °C contains about $4.2\times10^{22}$ atoms, but only $2.5\times10^{13}$ free electrons and $2.5\times10^{13}$ holes. The addition of 0.001% of arsenic (an impurity) donates an extra $10^{17}$ free electrons in the same volume and the electrical conductivity is increased by a factor of 10,000.

The materials chosen as suitable dopants depend on the atomic properties of both the dopant and the material to be doped. In general, dopants that produce the desired controlled changes are classified as either electron acceptors or donors. Semiconductors doped with *donor* impurities are called *n-type*, while those doped with *acceptor* impurities are known as *p-type*. The n and p type designations indicate which charge carrier acts as the material's majority carrier. The opposite carrier is called the minority carrier, which exists due to thermal excitation at a much lower concentration compared to the majority carrier.

For example, the pure semiconductor silicon has four valence electrons which bond each silicon atom to its neighbors. In silicon, the most common dopants are *group III* and *group V* elements. Group III elements all contain three valence electrons, causing them to function as acceptors when used to dope silicon. When an acceptor atom replaces a silicon atom in the crystal, a vacant state ( an electron "hole") is created, which can move around the lattice and functions as a charge carrier. Group V elements have five valence electrons, which allows them to act as a donor; substitution of these atoms for silicon creates an extra free electron. Therefore, a silicon crystal doped with boron creates a p-type semiconductor whereas one doped with phosphorus results in an n-type material.

During manufacture, dopants can be diffused into the semiconductor body by contact with gaseous compounds of the desired element, or ion implantation can be used to accurately position the doped regions.

## Early History of Semiconductors

The history of the understanding of semiconductors begins with experiments on the electrical properties of materials. The properties of negative temperature coefficient of resistance, rectification, and light-sensitivity were observed starting in the early 19th century.

Thomas Johann Seebeck was the first to notice an effect due to semiconductors, in 1821. In 1833, Michael Faraday reported that the resistance of specimens of silver sulfide decreases when they are heated. This is contrary to the behavior of metallic substances such as copper. In 1839, A. E. Becquerel reported observation of a voltage between a solid and a liquid electrolyte when struck by light, the photovoltaic effect. In 1873 Willoughby Smith observed that selenium resistors exhibit decreasing resistance when light falls on them. In 1874 Karl Ferdinand Braun observed conduction and rectification in metallic sulphides, although this effect had been discovered much earlier by M.A. Rosenschold writing for the Annalen der Physik und Chemie in 1835, and Arthur Schuster found that a copper oxide layer on wires has rectification properties that ceases when the wires are cleaned. Adams and Day observed the photovoltaic effect in selenium in 1876.

A unified explanation of these phenomena required a theory of solid-state physics which developed greatly in the first half of the 20th Century. In 1878 Edwin Herbert Hall demonstrated the deflection of flowing charge carriers by an applied magnetic field, the Hall effect. The discovery of the electron by J.J. Thomson in 1897 prompted theories of electron-based conduction in solids. Karl Baedeker, by observing a Hall effect with the reverse sign to that in metals, theorized that copper iodide had positive charge carriers. Johan Koenigsberger classified solid materials as metals, insulators and "variable conductors" in 1914 although his student Josef Weiss already introduced the term *Halbleiter* (semiconductor in modern meaning) in PhD thesis in 1910. Felix Bloch published a theory of the movement of electrons through atomic lattices in 1928. In 1930, B. Gudden stated that conductivity in semiconductors was due to minor concentrations of impurities. By 1931, the band theory of conduction had been established by Alan Herries Wilson and the concept of band gaps had been developed. Walter H. Schottky and Nevill Francis Mott developed models of the potential barrier and of the characteristics of a metal-semiconductor junction. By 1938, Boris Davydov had developed a theory of the copper-oxide rectifier, identifying the effect of the p–n junction and the importance of minority carriers and surface states.

Agreement between theoretical predictions (based on developing quantum mechanics) and experimental results was sometimes poor. This was later explained by John Bardeen as due to the extreme "structure sensitive" behavior of semiconductors, whose properties change dramatically based on tiny amounts of impurities. Commercially pure materials of the 1920s containing varying proportions of trace contaminants produced differing experimental results. This spurred the development of improved material refining techniques, culminating in modern semiconductor refineries producing materials with parts-per-trillion purity.

Devices using semiconductors were at first constructed based on empirical knowledge, before semiconductor theory provided a guide to construction of more capable and reliable devices.

Alexander Graham Bell used the light-sensitive property of selenium to transmit sound over a beam of light in 1880. A working solar cell, of low efficiency, was constructed by Charles Fritts in 1883 using a metal plate coated with selenium and a thin layer of gold; the device became commercially useful in photographic light meters in the 1930s. Point-contact microwave detector rectifiers made of lead sulfide were used by Jagadish Chandra Bose in 1904; the cat's-whisker detector using natural galena or other materials became a common device in the development of radio. However, it was somewhat unpredictable in operation and required manual adjustment for best performance. In 1906 H.J. Round observed light emission when electric current passed through silicon carbide crystals, the principle behind the light-emitting diode. Oleg Losev observed similar light emission in 1922 but at the time the effect had no practical use. Power rectifiers, using copper oxide and selenium, were developed in the 1920s and became commercially important as an alternative to vacuum tube rectifiers.

In the years preceding World War II, infra-red detection and communications devices prompted research into lead-sulfide and lead-selenide materials. These devices were used for detecting ships and aircraft, for infrared rangefinders, and for voice communication systems. The point-contact crystal detector became vital for microwave radio systems, since available vacuum tube devices could not serve as detectors above about 4000 MHz; advanced radar systems relied on the fast response of crystal detectors. Considerable research and development of silicon materials occurred during the war to develop detectors of consistent quality.

Detector and power rectifiers could not amplify a signal. Many efforts were made to develop a solid-state amplifier, but these were unsuccessful because of limited theoretical understanding of semiconductor materials. In 1922 Oleg Losev developed two-terminal, negative resistance amplifiers for radio; however, he perished in the Siege of Leningrad. In 1926 Julius Edgar Lilienfeld patented a device resembling a modern field-effect transistor, but it was not practical. R. Hilsch and R. W. Pohl in 1938 demonstrated a solid-state amplifier using a structure resembling the control grid of a vacuum tube; although the device displayed power gain, it had a cut-off frequency of one cycle per second, too low for any practical applications, but an effective application of the available theory. At Bell Labs, William Shockley and A. Holden started investigating solid-state amplifiers in 1938. The first p–n junction in silicon was observed by Russell Ohl about 1941, when a specimen was found to be light-sensitive, with a sharp boundary between p-type impurity at one end and n-type at the other. A slice cut from the specimen at the p–n boundary developed a voltage when exposed to light.

In France, during the war, Herbert Mataré had observed amplification between adjacent point contacts on a germanium base. After the war, Mataré's group announced their "Transistron" amplifier only shortly after Bell Labs announced the "transistor".

## Various Semiconductor Devices

### Diode

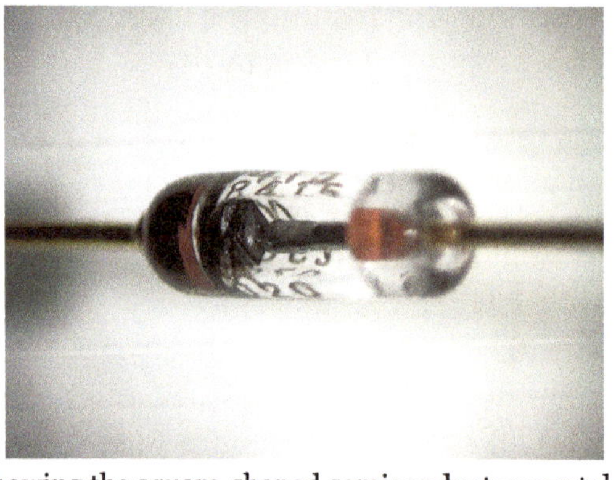

Closeup of a diode, showing the square-shaped semiconductor crystal *(black object on left)*

Extreme macro photo of a Chinese diode of the seventies

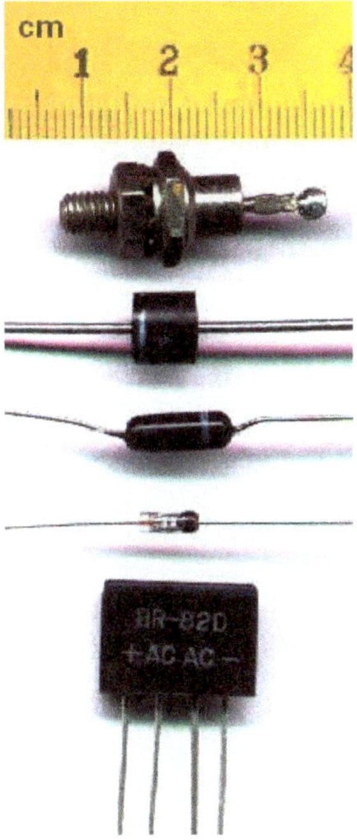


Various semiconductor diodes. Bottom: A bridge rectifier. In most diodes, a white or black painted band identifies the cathode into which electrons will flow when the diode is conducting. Electron flow is the reverse of conventional current flow

In electronics, a diode is a two-terminal electronic component that conducts primarily in one direction (asymmetric conductance); it has low (ideally zero) resistance to the current in one direction, and high (ideally infinite) resistance in the other. A semiconductor diode, the most common type today, is a crystalline piece of semiconductor material with a p–n junction connected to two electrical terminals. A vacuum tube diode has two electrodes, a plate (anode) and a heated cathode. Semiconductor diodes were the first semiconductor electronic devices. The discovery of crystals' rectifying abilities was made by German physicist Ferdinand Braun in 1874. The first semiconductor diodes, called cat's whisker diodes, developed around 1906, were made of mineral crystals such as galena. Today, most diodes are made of silicon, but other semiconductors such as selenium and germanium are sometimes used.

## Main Functions

The most common function of a diode is to allow an electric current to pass in one direction (called the diode's *forward* direction), while blocking current in the opposite direction (the *reverse* direction). Thus, the diode can be viewed as an electronic version of a check valve. This unidirectional behavior is called rectification, and is used to convert alternating current (AC) to direct current (DC), including extraction of modulation from radio signals in radio receivers—these diodes are forms of rectifiers.

However, diodes can have more complicated behavior than this simple on–off action, because of their nonlinear current-voltage characteristics. Semiconductor diodes begin conducting electricity only if a certain threshold voltage or cut-in voltage is present in the forward direction (a state in which the diode is said to be *forward-biased*). The voltage drop across a forward-biased diode varies only a little with the current, and is a function of temperature; this effect can be used as a temperature sensor or as a voltage reference.

A semiconductor diode's current–voltage characteristic can be tailored by selecting the semiconductor materials and the doping impurities introduced into the materials during manufacture. These techniques are used to create special-purpose diodes that perform many different functions. For example, diodes are used to regulate voltage (Zener diodes), to protect circuits from high voltage surges (avalanche diodes), to electronically tune radio and TV receivers (varactor diodes), to generate radio-frequency oscillations (tunnel diodes, Gunn diodes, IMPATT diodes), and to produce light (light-emitting diodes). Tunnel, Gunn and IMPATT diodes exhibit negative resistance, which is useful in microwave and switching circuits.

Diodes, both vacuum and semiconductor, can be used as shot-noise generators.

## History

Thermionic (vacuum tube) diodes and solid state (semiconductor) diodes were developed separately, at approximately the same time, in the early 1900s, as radio receiver detectors. Until the 1950s vacuum tube diodes were used more frequently in radios because the early point-contact type semiconductor diodes were less stable. In addition, most receiving sets had vacuum tubes for amplification that could easily have the thermionic diodes included in the tube (for example the 12SQ7 double diode triode), and vacuum tube rectifiers and gas-filled rectifiers were capable of handling some high voltage/high current rectification tasks better than the semiconductor diodes (such as selenium rectifiers) which were available at that time.

## Vacuum Tube Diodes

In 1873, Frederick Guthrie discovered the basic principle of operation of thermionic diodes. Guthrie discovered that a positively charged electroscope could be discharged by bringing a grounded piece of white-hot metal close to it (but not actually touching it). The same did not apply to a negatively charged electroscope, indicating that the current flow was only possible in one direction.

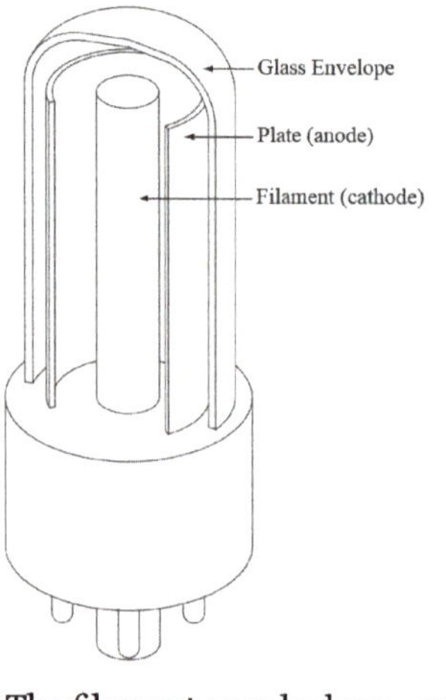


Structure of a vacuum tube diode. The filament may be bare, or more commonly (as shown here), embedded within and insulated from an enclosing cathode

Thomas Edison independently rediscovered the principle on February 13, 1880. At the time, Edison was investigating why the filaments of his carbon-filament light bulbs nearly always burned out at the positive-connected end. He had a special bulb made with a metal plate sealed into the glass envelope. Using this device, he confirmed that an invisible current flowed from the glowing filament through the vacuum to the metal plate, but only when the plate was connected to the positive supply.

Edison devised a circuit where his modified light bulb effectively replaced the resistor in a DC voltmeter. Edison was awarded a patent for this invention in 1884. Since there was no apparent practical use for such a device at the time, the patent application was most likely simply a precaution in case someone else did find a use for the so-called Edison effect.

About 20 years later, John Ambrose Fleming (scientific adviser to the Marconi Company and former Edison employee) realized that the Edison effect could be used as a precision radio detector. Fleming patented the first true thermionic diode, the Fleming valve, in Britain on November 16, 1904 (followed by U.S. Patent 803,684 in November 1905).

## Solid-state Diodes

In 1874 German scientist Karl Ferdinand Braun discovered the "unilateral conduction" of crystals. Braun patented the crystal rectifier in 1899. Copper oxide and selenium rectifiers were developed for power applications in the 1930s.

Indian scientist Jagadish Chandra Bose was the first to use a crystal for detecting radio waves in 1894. The crystal detector was developed into a practical device for wireless telegraphy by Greenleaf Whittier Pickard, who invented a silicon crystal detector in 1903 and received a patent for it on November 20, 1906. Other experimenters tried a variety of other substances, of which the most widely used was the mineral galena (lead sulfide). Other substances offered slightly better performance, but galena was most widely used because it had the advantage of being cheap and easy to obtain. The crystal detector in these early crystal radio sets consisted of an adjustable wire point-contact, often made of gold or platinum because of their incorrodible nature (the so-called "cat's whisker"), which could be manually moved over the face of the crystal in search of a portion of that mineral with rectifying qualties. This troublesome device was superseded by thermionic diodes (vacuum tubes) by the 1920s, but after high purity semiconductor materials became available, the crystal detector returned to dominant use with the advent, in the 1950s, of inexpensive fixed-germanium diodes. Bell Labs also developed a germanium diode for microwave reception, and AT&T used these in their microwave towers that criss-crossed the nation starting in the late 1940s, carrying telephone and network television signals. Bell Labs did not develop a satisfactory thermionic diode for microwave reception.

## Etymology

At the time of their invention, such devices were known as rectifiers. In 1919, the year tetrodes were invented, William Henry Eccles coined the term diode from the Greek roots di, meaning 'two', and ode, meaning 'path'. (However, the word *diode* itself, as well as *triode, tetrode, pentode, hexode*, were already in use as terms of multiplex telegraphy.

## Rectifiers

Although all diodes *rectify*, the term 'rectifier' is normally reserved for higher currents and voltages than would normally be found in the rectification of lower power signals; examples include:

- Power supply rectifiers (*half-wave, full-wave, bridge*)
- Flyback diodes

## World's Smallest Diode

Researchers from the University of Georgia and Ben-Gurion University of the Negev (BGU) have developed a diode made from a molecule of DNA. Professor Bingqian Xu from the College of Engineering at the University of Georgia and his team took a single DNA molecule made from 11 base pairs and connected it to an electronic circuit a few nanometers in size. When layers of coralyne were inserted between layers of DNA, the current jumped up to 15 times larger negative versus positive, which is necessary for a nano diode.

## Thermionic Diodes

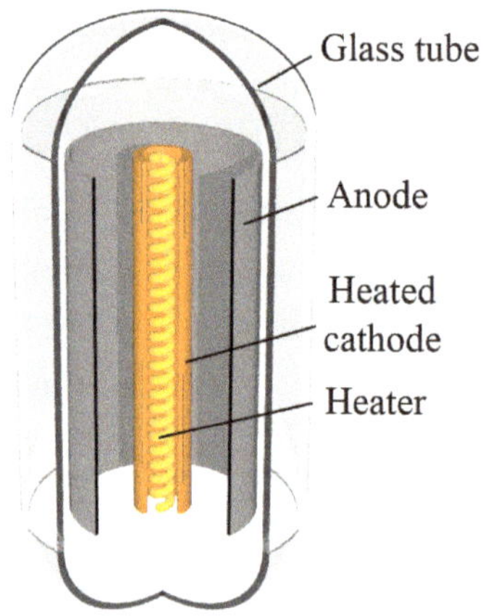


Diode vacuum tube construction

A thermionic diode is a thermionic-valve device (also known as a vacuum tube, tube, or valve), consisting of a sealed evacuated glass envelope containing two electrodes: a cathode heated by a filament, and a plate (anode). Early examples were fairly similar in appearance to incandescent light bulbs.

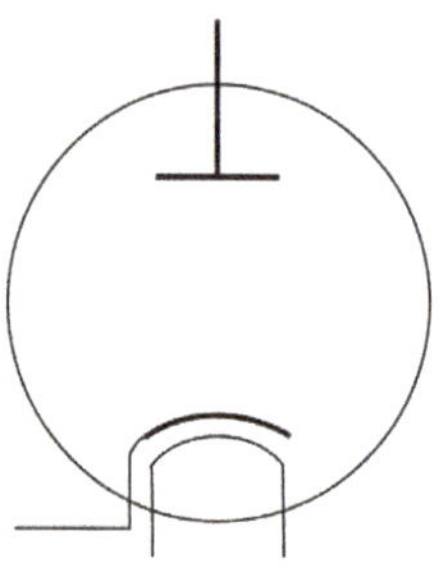

The symbol for an indirect heated vacuum-tube diode. From top to bottom, the components are the anode, the cathode, and the heater filament

In operation, a current flows through the filament (heater)—a high resistance wire made of nichrome—and heats the cathode red hot (800–1000 °C). This causes the cathode to release electrons into the vacuum, a process called thermionic emission. (Some valves use *direct heating*, in which a tungsten filament acts as both heater and cathode.) The alternating voltage to be rectified is applied between the cathode and the concentric plate electrode. When the plate has a positive voltage with respect to the cathode, it electrostatically attracts the electrons from the cathode, so a current of electrons flows through the tube from cathode to plate. However, when the polarity is reversed and the plate has a negative voltage, no current flows, because the cathode electrons are not attracted to it. The plate, being unheated, does not emit any electrons. So electrons can only flow through the tube in one direction, from the cathode to the anode plate.

The cathode is coated with oxides of alkaline earth metals, such as barium and strontium oxides. These have a low work function, meaning that they more readily emit electrons than would the uncoated cathode.

In a mercury-arc valve, an arc forms between a refractory conductive anode and a pool of liquid mercury acting as cathode. Such units were made with ratings up to hundreds of kilowatts, and were important in the development of HVDC power transmission. Some types of smaller thermionic rectifiers had mercury vapor fill to reduce their forward voltage drop and to increase current rating over thermionic hard-vacuum devices.

Throughout the vacuum tube era, valve diodes were used in analog signal applications and as rectifiers in DC power supplies in consumer electronics such as radios, televisions, and sound systems. They were replaced in power supplies beginning in the 1940s by selenium rectifiers and then by semiconductor diodes by the 1960s. Today they are still used in a few high power applications where their ability to withstand transient voltages and their robustness gives them an advantage over semiconductor devices. The recent (2012) resurgence of interest among audiophiles and recording studios in old valve audio gear such as guitar amplifiers and home audio systems has provided a market for the legacy consumer diode valves.

## Semiconductor Diodes

### Electronic Symbols

The symbol used for a semiconductor diode in a circuit diagram specifies the type of diode. There are alternative symbols for some types of diodes, though the differences are minor. The triangle in the symbols points to the forward direction, i.e. in the direction of current flow.

| | | |
|---|---|---|
|  | 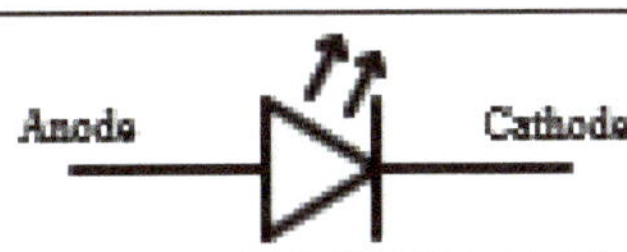 | 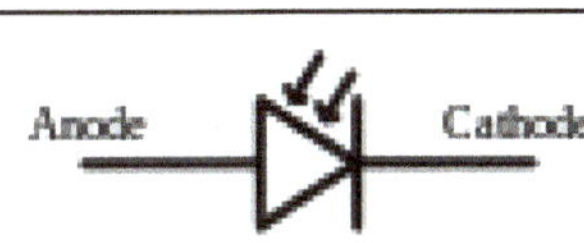 |
| Diode | Light-emitting diode (LED) | Photodiode |
| 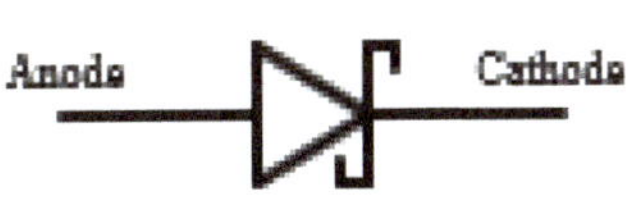 |  | 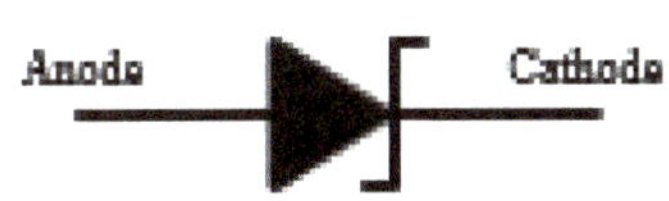 |
| Schottky diode | Transient-voltage-suppression diode (TVS) | Tunnel diode |

<table>
<tr>
<td></td>
<td>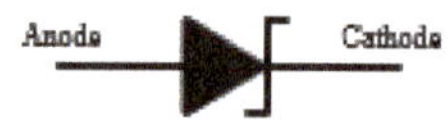</td>
<td>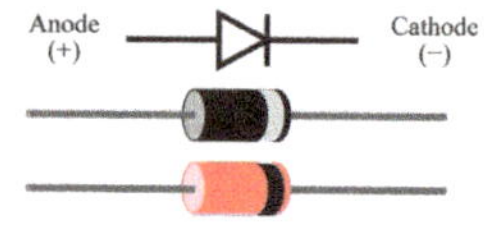</td>
</tr>
<tr>
<td>Varicap</td>
<td>Zener diode</td>
<td>Typical diode packages in same alignment as diode symbol. Thin bar depicts the cathode.</td>
</tr>
</table>

A galena cat's-whisker detector, a point-contact diode.

## Point-contact Diodes

A point-contact diode works the same as the junction diodes described below, but its construction is simpler. A pointed metal wire is placed in contact with an n-type semiconductor. Some metal migrates into the semiconductor to make a small p-type region around the contact. The 1N34 germanium version is still used in radio receivers as a detector and occasionally in specialized analog electronics.

## Junction Diodes

### p–n junction Diode

A p–n junction diode is made of a crystal of semiconductor, usually silicon, but germanium and gallium arsenide are also used. Impurities are added to it to create a region on one side that contains negative charge carriers (electrons), called an n-type semiconductor, and a region on the other side that contains positive charge carriers (holes), called a p-type semiconductor. When the n-type and p-type materials are attached together, a momentary flow of electrons occur from the n to the p side resulting in a third region between the two where no charge carriers are present. This region is called the depletion region because there are no charge carriers (neither electrons nor holes) in it. The diode's terminals are attached to the n-type and p-regions. The boundary between these two regions, called a p–n junction, is where the action of the diode takes place. When a sufficiently higher electrical potential is applied to the P side (the anode) than to the N side (the cathode), it allows electrons to flow through the depletion region from the N-type side to the P-type side. The junction does not allow the flow of electrons in the opposite direction when the potential is applied in reverse, creating, in a sense, an electrical check valve.

## Schottky Diode

Another type of junction diode, the Schottky diode, is formed from a metal–semiconductor junction rather than a p–n junction, which reduces capacitance and increases switching speed.

## Current–voltage Characteristic

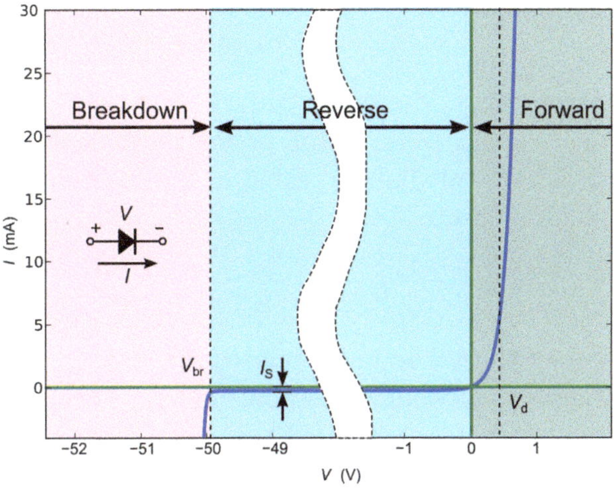


I–V (current vs. voltage) characteristics of a p–n junction diode

A semiconductor diode's behavior in a circuit is given by its current–voltage characteristic, or I–V graph. The shape of the curve is determined by the transport of charge carriers through the so-called *depletion layer* or *depletion region* that exists at the p–n junction between differing semiconductors. When a p–n junction is first created, conduction-band (mobile) electrons from the N-doped region diffuse into the P-doped region where there is a large population of holes (vacant places for electrons) with which the electrons "recombine". When a mobile electron recombines with a hole, both hole and electron vanish, leaving behind an immobile positively charged donor (dopant) on the N side and negatively charged acceptor (dopant) on the P side. The region around the p–n junction becomes depleted of charge carriers and thus behaves as an insulator.

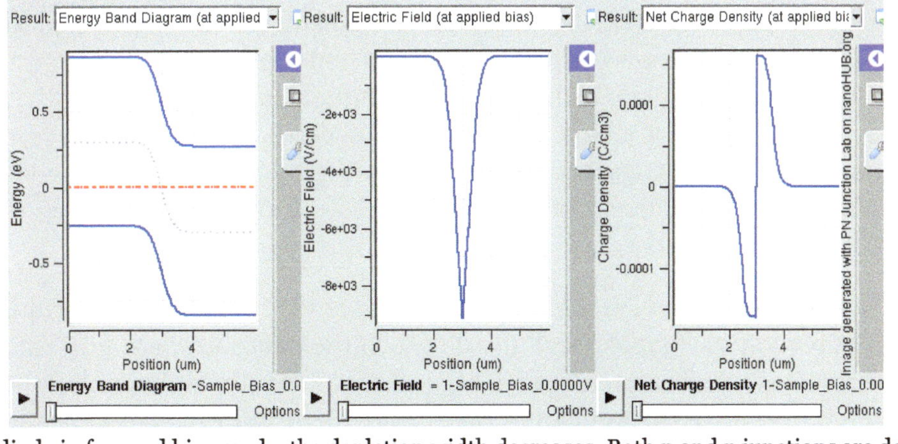


A PN junction diode in forward bias mode, the depletion width decreases. Both p and n junctions are doped at a 1e15/cm3 doping level, leading to built-in potential of ~0.59V. Observe the different Quasi Fermi levels for conduction band and valence band in n and p regions (red curves)

However, the width of the depletion region (called the depletion width) cannot grow without limit. For each electron–hole pair recombination made, a positively charged dopant ion is left behind in the N-doped region, and a negatively charged dopant ion is created in the P-doped region. As recombination proceeds and more ions are created, an increasing electric field develops through the depletion zone that acts to slow and then finally stop recombination. At this point, there is a "built-in" potential across the depletion zone.

## Reverse Bias

If an external voltage is placed across the diode with the same polarity as the built-in potential, the depletion zone continues to act as an insulator, preventing any significant electric current flow (unless electron–hole pairs are actively being created in the junction by, for instance, light). This is called the *reverse bias* phenomenon.

## Forward Bias

However, if the polarity of the external voltage opposes the built-in potential, recombination can once again proceed, resulting in a substantial electric current through the p–n junction (i.e. substantial numbers of electrons and holes recombine at the junction). For silicon diodes, the built-in potential is approximately 0.7 V (0.3 V for germanium and 0.2 V for Schottky). Thus, if an external voltage greater than and opposite to the built-in voltage is applied, a current will flow and the diode is said to be "turned on" as it has been given an external *forward bias*. The diode is commonly said to have a forward "threshold" voltage, above which it conducts and below which conduction stops. However, this is only an approximation as the forward characteristic is according to the Shockley equation absolutely smooth.

A diode's I–V characteristic can be approximated by four regions of operation:

1. At very large reverse bias, beyond the peak inverse voltage or PIV, a process called reverse breakdown occurs that causes a large increase in current (i.e., a large number of electrons and holes are created at, and move away from the p–n junction) that usually damages the device permanently. The avalanche diode is deliberately designed for use in that manner. In the Zener diode, the concept of PIV is not applicable. A Zener diode contains a heavily doped p–n junction allowing electrons to tunnel from the valence band of the p-type material to the conduction band of the n-type material, such that the reverse voltage is "clamped" to a known value (called the *Zener voltage*), and avalanche does not occur. Both devices, however, do have a limit to the maximum current and power they can withstand in the clamped reverse-voltage region. Also, following the end of forward conduction in any diode, there is reverse current for a short time. The device does not attain its full blocking capability until the reverse current ceases.

2. For a bias less than the PIV, the reverse current is very small. For a normal P–N rectifier diode, the reverse current through the device in the micro-ampere (μA) range is very low. However, this is temperature dependent, and at sufficiently high temperatures, a substantial amount of reverse current can be observed (mA or more).

3. With a small forward bias, where only a small forward current is conducted, the current–

voltage curve is exponential in accordance with the ideal diode equation. There is a definite forward voltage at which the diode starts to conduct significantly. This is called the *knee voltage* or *cut-in voltage* and is equal to the barrier potential of the p-n junction. This is a feature of the exponential curve, and appears sharper on a current scale more compressed than in the diagram shown here.

4. At larger forward currents the current-voltage curve starts to be dominated by the ohmic resistance of the bulk semiconductor. The curve is no longer exponential, it is asymptotic to a straight line whose slope is the bulk resistance. This region is particularly important for power diodes. The diode can be modeled as an ideal diode in series with a fixed resistor.

In a small silicon diode operating at its rated currents, the voltage drop is about 0.6 to 0.7 volts. The value is different for other diode types—Schottky diodes can be rated as low as 0.2 V, germanium diodes 0.25 to 0.3 V, and red or blue light-emitting diodes (LEDs) can have values of 1.4 V and 4.0 V respectively.

At higher currents the forward voltage drop of the diode increases. A drop of 1 V to 1.5 V is typical at full rated current for power diodes.

## Shockley Diode Equation

The *Shockley ideal diode equation* or the *diode law* (named after the bipolar junction transistor co-inventor William Bradford Shockley) gives the I–V characteristic of an ideal diode in either forward or reverse bias (or no bias). The following equation is called the *Shockley ideal diode equation* when *n*, the ideality factor, is set equal to 1 :

$$I = I_S \left( e^{\frac{V_D}{nV_T}} - 1 \right)$$

where

$I$ is the diode current,

$I_S$ is the reverse bias saturation current (or scale current),

$V_D$ is the voltage across the diode,

$V_T$ is the thermal voltage, and

$n$ is the *ideality factor*, also known as the *quality factor* or sometimes *emission coefficient*. The ideality factor $n$ typically varies from 1 to 2 (though can in some cases be higher), depending on the fabrication process and semiconductor material and is set equal to 1 for the case of an "ideal" diode (thus the n is sometimes omitted). The ideality factor was added to account for imperfect junctions as observed in real transistors. The factor mainly accounts for carrier recombination as the charge carriers cross the depletion region.

The thermal voltage $V_T$ is approximately 25.85 mV at 300 K, a temperature close to "room temperature" commonly used in device simulation software. At any temperature it is a known constant defined by:

$$V_T = \frac{kT}{q},$$

where $k$ is the Boltzmann constant, $T$ is the absolute temperature of the p–n junction, and $q$ is the magnitude of charge of an electron (the elementary charge).

The reverse saturation current, $I_S$, is not constant for a given device, but varies with temperature; usually more significantly than $V_T$, so that $V_D$ typically decreases as $T$ increases.

The *Shockley ideal diode equation* or the *diode law* is derived with the assumption that the only processes giving rise to the current in the diode are drift (due to electrical field), diffusion, and thermal recombination–generation (R–G) (this equation is derived by setting n = 1 above). It also assumes that the R–G current in the depletion region is insignificant. This means that the *Shockley ideal diode equation* doesn't account for the processes involved in reverse breakdown and photon-assisted R–G. Additionally, it doesn't describe the "leveling off" of the I–V curve at high forward bias due to internal resistance. Introducing the ideality factor, n, accounts for recombination and generation of carriers.

Under *reverse bias* voltages the exponential in the diode equation is negligible, and the current is a constant (negative) reverse current value of $-I_S$. The reverse *breakdown region* is not modeled by the Shockley diode equation.

For even rather small *forward bias* voltages the exponential is very large, since the thermal voltage is very small in comparison. The subtracted '1' in the diode equation is then negligible and the forward diode current can be approximated by

$$I = I_S e^{\frac{V_D}{nV_T}}$$

## Small-signal Behavior

For circuit design, a small-signal model of the diode behavior often proves useful. A specific example of diode modeling is discussed in the article on small-signal circuits.

## Reverse-recovery Effect

Following the end of forward conduction in a p–n type diode, a reverse current can flow for a short time. The device does not attain its blocking capability until the mobile charge in the junction is depleted.

The effect can be significant when switching large currents very quickly. A certain amount of "reverse recovery time" $t_r$ (on the order of tens of nanoseconds to a few microseconds) may be required to remove the reverse recovery charge $Q_r$ from the diode. During this recovery time, the diode can actually conduct in the reverse direction. This might give rise to a large constant current in the reverse direction for a short time while the diode is reverse biased. The magnitude of such a reverse current is determined by the operating circuit (i.e., the series resistance) and the diode is said to be in the storage-phase. In certain real-world cases it is important to consider the losses that are incurred by this non-ideal diode effect. However, when the slew rate of the current is not so severe (e.g. Line frequency) the effect can be safely ignored. For most applications, the effect is also negligible for Schottky diodes.

The reverse current ceases abruptly when the stored charge is depleted; this abrupt stop is exploited in step recovery diodes for generation of extremely short pulses.

## Types of Semiconductor Diode

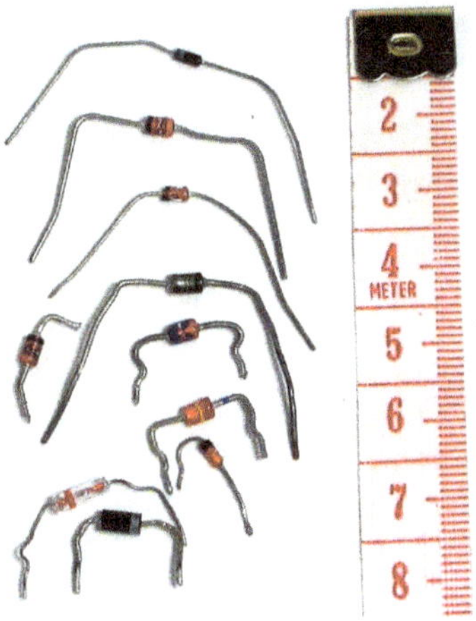


Several types of diodes. The scale is centimeters

There are several types of p–n junction diodes, which emphasize either a different physical aspect of a diode often by geometric scaling, doping level, choosing the right electrodes, are just an application of a diode in a special circuit, or are really different devices like the Gunn and laser diode and the MOSFET:

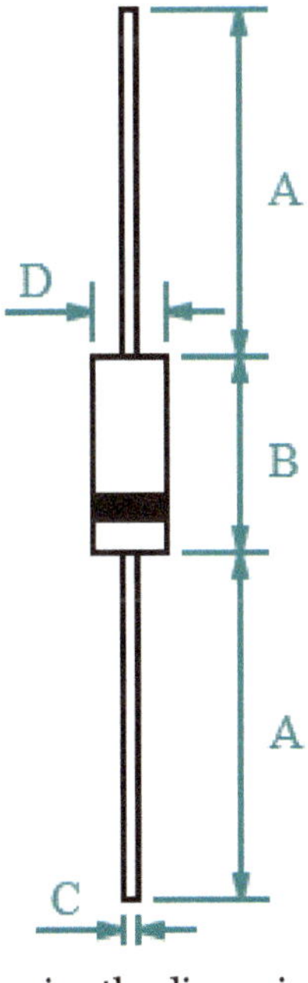


Typical datasheet drawing showing the dimensions of a DO-41 diode package

Normal (p–n) diodes, which operate as described above, are usually made of doped silicon or, more rarely, germanium. Before the development of silicon power rectifier diodes, cuprous oxide and later selenium was used. Their low efficiency required a much higher forward voltage to be applied (typically 1.4 to 1.7 V per "cell", with multiple cells stacked so as to increase the peak inverse voltage rating for application in high voltage rectifiers), and required a large heat sink (often an extension of the diode's metal substrate), much larger than the later silicon diode of the same current ratings would require. The vast majority of all diodes are the p–n diodes found in CMOS integrated circuits, which include two diodes per pin and many other internal diodes.

Avalanche diodes

These are diodes that conduct in the reverse direction when the reverse bias voltage exceeds the breakdown voltage. These are electrically very similar to Zener diodes (and are often mistakenly called Zener diodes), but break down by a different mechanism: the *avalanche effect*. This occurs when the reverse electric field applied across the p–n junction causes a wave of ionization, reminiscent of an avalanche, leading to a large current. Avalanche diodes are designed to break down at a well-defined reverse voltage without being destroyed. The difference between the avalanche diode (which has a reverse breakdown above about 6.2 V) and the Zener is that the channel length of the former exceeds the mean free path of the electrons, resulting in many collisions between them on the way through the channel. The only practical difference between the two types is they have temperature coefficients of opposite polarities.

Cat's whisker or crystal diodes

These are a type of point-contact diode. The cat's whisker diode consists of a thin or sharpened metal wire pressed against a semiconducting crystal, typically galena or a piece of coal. The wire forms the anode and the crystal forms the cathode. Cat's whisker diodes were also called crystal diodes and found application in the earliest radios called crystal radio receivers. Cat's whisker diodes are generally obsolete, but may be available from a few manufacturers.

Constant current diodes

These are actually JFETs with the gate shorted to the source, and function like a two-terminal current-limiting analog to the voltage-limiting Zener diode. They allow a current through them to rise to a certain value, and then level off at a specific value. Also called *CLDs, constant-current diodes, diode-connected transistors*, or *current-regulating diodes*.

Esaki or tunnel diodes

These have a region of operation showing negative resistance caused by quantum tunneling, allowing amplification of signals and very simple bistable circuits. Because of the high carrier concentration, tunnel diodes are very fast, may be used at low (mK) temperatures, high magnetic fields, and in high radiation environments. Because of these properties, they are often used in spacecraft.

Gunn diodes

These are similar to tunnel diodes in that they are made of materials such as GaAs or InP that exhibit a region of negative differential resistance. With appropriate biasing, dipole domains form and travel across the diode, allowing high frequency microwave oscillators to be built.

Light-emitting diodes (LEDs)

In a diode formed from a direct band-gap semiconductor, such as gallium arsenide, charge carriers that cross the junction emit photons when they recombine with the majority carrier on the other side. Depending on the material, wavelengths (or colors) from the infrared to the near ultraviolet may be produced. The forward potential of these diodes depends on the wavelength of the emitted photons: 2.1 V corresponds to red, 4.0 V to violet. The first LEDs were red and yellow, and higher-frequency diodes have been developed over time. All LEDs produce incoherent, narrow-spectrum light; "white" LEDs are actually combinations of three LEDs of a different color, or a blue LED with a yellow scintillator coating. LEDs can also be used as low-efficiency photodiodes in signal applications. An LED may be paired with a photodiode or phototransistor in the same package, to form an opto-isolator.

Laser diodes

When an LED-like structure is contained in a resonant cavity formed by polishing the parallel end faces, a laser can be formed. Laser diodes are commonly used in optical storage devices and for high speed optical communication.

Thermal diodes

This term is used both for conventional p–n diodes used to monitor temperature because of their varying forward voltage with temperature, and for Peltier heat pumps for thermoelectric heating and cooling. Peltier heat pumps may be made from semiconductor, though they do not have any rectifying junctions, they use the differing behaviour of charge carriers in N and P type semiconductor to move heat.

Perun's diodes

This is a special type of voltage-surge protection diode. It is characterized by the symmetrical voltage-current characteristic, similar to DIAC. It has much faster response time however, that's why it is used in demanding applications.

Photodiodes

All semiconductors are subject to optical charge carrier generation. This is typically an undesired effect, so most semiconductors are packaged in light blocking material. Photodiodes are intended to sense light(photodetector), so they are packaged in materials that allow light to pass, and are usually PIN (the kind of diode most sensitive to light). A photodiode can be used in solar cells, in photometry, or in optical communications. Multiple photodiodes may be packaged in a single device, either as a linear array or as a two-dimensional array. These arrays should not be confused with charge-coupled devices.

PIN diodes

A PIN diode has a central un-doped, or *intrinsic*, layer, forming a p-type/intrinsic/n-type structure. They are used as radio frequency switches and attenuators. They are also used as large-volume, ionizing-radiation detectors and as photodetectors. PIN diodes are also used in power electronics, as their central layer can withstand high voltages. Furthermore, the PIN structure can be found in many power semiconductor devices, such as IGBTs, power MOSFETs, and thyristors.

Schottky diodes

Schottky diodes are constructed from a metal to semiconductor contact. They have a lower forward voltage drop than p–n junction diodes. Their forward voltage drop at forward currents of about 1 mA is in the range 0.15 V to 0.45 V, which makes them useful in voltage clamping applications and prevention of transistor saturation. They can also be used as low loss rectifiers, although their reverse leakage current is in general higher than that of other diodes. Schottky diodes are majority carrier devices and so do not suffer from minority carrier storage problems that slow down many other diodes—so they have a faster reverse recovery than p–n junction diodes. They also tend to have much lower junction capacitance than p–n diodes, which provides for high switching speeds and their use in high-speed circuitry and RF devices such as switched-mode power supply, mixers, and detectors.

Super barrier diodes

Super barrier diodes are rectifier diodes that incorporate the low forward voltage drop of the Schottky diode with the surge-handling capability and low reverse leakage current of a normal p–n junction diode.

Gold-doped diodes

As a dopant, gold (or platinum) acts as recombination centers, which helps a fast recombination of minority carriers. This allows the diode to operate at signal frequencies, at the expense of a higher forward voltage drop. Gold-doped diodes are faster than other p–n diodes (but not as fast as Schottky diodes). They also have less reverse-current leakage than Schottky diodes (but not as good as other p–n diodes). A typical example is the 1N914.

Snap-off or Step recovery diodes

The term *step recovery* relates to the form of the reverse recovery characteristic of these devices. After a forward current has been passing in an SRD and the current is interrupted or reversed, the reverse conduction will cease very abruptly (as in a step waveform). SRDs can, therefore, provide very fast voltage transitions by the very sudden disappearance of the charge carriers.

Stabistors or *Forward Reference Diodes*

The term *stabistor* refers to a special type of diodes featuring extremely stable forward voltage characteristics. These devices are specially designed for low-voltage stabilization applications requiring a guaranteed voltage over a wide current range and highly stable over temperature.

Transient voltage suppression diode (TVS)

These are avalanche diodes designed specifically to protect other semiconductor devices from high-voltage transients. Their p–n junctions have a much larger cross-sectional area than those of a normal diode, allowing them to conduct large currents to ground without sustaining damage.

Varicap or varactor diodes

These are used as voltage-controlled capacitors. These are important in PLL (phase-locked loop) and FLL (frequency-locked loop) circuits, allowing tuning circuits, such as those in television receivers, to lock quickly on to the frequency. They also enabled tunable oscillators in early discrete tuning of radios, where a cheap and stable, but fixed-frequency, crystal oscillator provided the reference frequency for a voltage-controlled oscillator.

Zener diodes

These can be made to conduct in reverse bias (backward), and are correctly termed reverse breakdown diodes. This effect, called Zener breakdown, occurs at a precisely defined voltage, allowing the diode to be used as a precision voltage reference. The term Zener diode is colloquially applied to several types of breakdown diodes, but strictly speaking Zener diodes have a breakdown voltage of below 5 volts, whilst avalanche diodes are used for breakdown voltages above that value. In practical voltage reference circuits, Zener and switching diodes are connected in series and opposite directions to balance the temperature coefficient response of the diodes to near-zero. Some devices labeled as high-voltage Zener diodes are actually avalanche diodes. Two (equivalent) Zeners in series and in reverse order, in the same package, constitute a transient absorber (or Transorb, a registered trademark).

Other uses for semiconductor diodes include the sensing of temperature, and computing analog logarithms.

## Numbering and Coding Schemes

There are a number of common, standard and manufacturer-driven numbering and coding schemes for diodes; the two most common being the EIA/JEDEC standard and the European Pro Electron standard:

### EIA/JEDEC

The standardized 1N-series numbering *EIA370* system was introduced in the US by EIA/JEDEC (Joint Electron Device Engineering Council) about 1960. Most diodes have a 1-prefix designation (e.g., 1N4003). Among the most popular in this series were: 1N34A/1N270 (germanium signal), 1N914/1N4148 (silicon signal), 1N400x (silicon 1A power rectifier), and 1N580x (silicon 3A power rectifier).

### JIS

The JIS semiconductor designation system has all semiconductor diode designations starting with “1S”.

## Pro Electron

The European Pro Electron coding system for active components was introduced in 1966 and comprises two letters followed by the part code. The first letter represents the semiconductor material used for the component (A = germanium and B = silicon) and the second letter represents the general function of the part (for diodes, A = low-power/signal, B = variable capacitance, X = multiplier, Y = rectifier and Z = voltage reference); for example:

- AA-series germanium low-power/signal diodes (e.g., AA119)
- BA-series silicon low-power/signal diodes (e.g., BAT18 silicon RF switching diode)
- BY-series silicon rectifier diodes (e.g., BY127 1250V, 1A rectifier diode)
- BZ-series silicon Zener diodes (e.g., BZY88C4V7 4.7V Zener diode)

Other common numbering / coding systems (generally manufacturer-driven) include:

- GD-series germanium diodes (e.g., GD9) – this is a very old coding system
- OA-series germanium diodes (e.g., OA47) – a coding sequence developed by Mullard, a UK company

As well as these common codes, many manufacturers or organisations have their own systems too – for example:

- HP diode 1901-0044 = JEDEC 1N4148
- UK military diode CV448 = Mullard type OA81 = GEC type GEX23

## Transistor

Assorted discrete transistors. Packages in order from top to bottom: TO-3, TO-126, TO-92, SOT-23

A transistor is a semiconductor device used to amplify or switch electronic signals and electrical power. It is composed of semiconductor material usually with at least three terminals for connection to an external circuit. A voltage or current applied to one pair of the transistor's terminals controls the current through another pair of terminals. Because the controlled (output) power can be higher than the controlling (input) power, a transistor can amplify a signal. Today, some transistors are packaged individually, but many more are found embedded in integrated circuits.

The transistor is the fundamental building block of modern electronic devices, and is ubiquitous in modern electronic systems. Julius Edgar Lilienfeld patented a field-effect transistor in 1926 but it was not possible to actually construct a working device at that time. The first practically implemented device was a point-contact transistor invented in 1947 by American physicists John Bardeen, Walter Brattain, and William Shockley. The transistor revolutionized the field of electronics, and paved the way for smaller and cheaper radios, calculators, and computers, among other things. The transistor is on the list of IEEE milestones in electronics, and Bardeen, Brattain, and Shockley shared the 1956 Nobel Prize in Physics for their achievement.

## History

The thermionic triode, a vacuum tube invented in 1907, enabled amplified radio technology and long-distance telephony. The triode, however, was a fragile device that consumed a substantial amount of power. In 1909 physicist William Eccles discovered the crystal diode oscillator. Physicist Julius Edgar Lilienfeld filed a patent for a field-effect transistor (FET) in Canada in 1925, which was intended to be a solid-state replacement for the triode. Lilienfeld also filed identical patents in the United States in 1926 and 1928. However, Lilienfeld did not publish any research articles about his devices nor did his patents cite any specific examples of a working prototype. Because the production of high-quality semiconductor materials was still decades away, Lilienfeld's solid-state amplifier ideas would not have found practical use in the 1920s and 1930s, even if such a device had been built. In 1934, German inventor Oskar Heil patented a similar device in Europe.

John Bardeen, William Shockley and Walter Brattain at Bell Labs, 1948

From November 17, 1947 to December 23, 1947, John Bardeen and Walter Brattain at AT&T's Bell Labs in Murray Hill, New Jersey of the United States performed experiments and observed that when two gold point contacts were applied to a crystal of germanium, a signal was produced with the output power greater than the input. Solid State Physics Group leader William Shockley saw the potential in this, and over the next few months worked to greatly expand the knowledge of semiconductors. The term *transistor* was coined by John R. Pierce as a contraction of the term *transresistance*. According to Lillian Hoddeson and Vicki Daitch, authors of a biography of John Bardeen, Shockley had proposed that Bell Labs' first patent for a transistor should be based on the field-effect and that he be named as the inventor. Having unearthed Lilienfeld's patents that went into obscurity years earlier, lawyers at Bell Labs advised against Shockley's proposal because the idea of a field-effect transistor that used an electric field as a "grid" was not new. Instead, what Bardeen, Brattain, and Shockley invented in 1947 was the first point-contact transistor. In acknowledgement of this accomplishment, Shockley, Bardeen, and Brattain were jointly awarded the 1956 Nobel Prize in Physics "for their researches on semiconductors and their discovery of the transistor effect".

Herbert F. Mataré (1950)

In 1948, the point-contact transistor was independently invented by German physicists Herbert Mataré and Heinrich Welker while working at the *Compagnie des Freins et Signaux*, a Westinghouse subsidiary located in Paris. Mataré had previous experience in developing crystal rectifiers from silicon and germanium in the German radar effort during World War II. Using this knowledge, he began researching the phenomenon of "interference" in 1947. By June 1948, witnessing currents flowing through point-contacts, Mataré produced consistent results using samples of germanium produced by Welker, similar to what Bardeen and Brattain had accomplished earlier in December 1947. Realizing that Bell Labs' scientists had already invented the transistor before them, the company rushed to get its "transistron" into production for amplified use in France's telephone network.

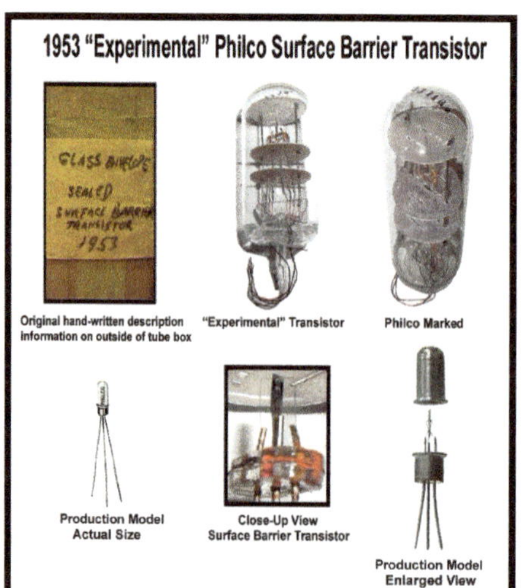


Philco surface-barrier transistor developed and produced in 1953

The first bipolar junction transistors were invented by Bell labs William Shockley, which applied for patent(2,569,347) on June 26, 1948. On April 12, 1950, Bell Labs chemists Gordon Teal and Morgan Sparks had successfully produced a working bipolar NPN junction amplifying germanium transistor. Bell Labs had made this new "sandwich" transistor discovery announcement, in a press release on July 4, 1951.

The first high-frequency transistor was the surface-barrier germanium transistor developed by Philco in 1953, capable of operating up to 60 MHz. These were made by etching depressions into an N-type germanium base from both sides with jets of Indium(III) sulfate until it was a few ten-thousandths of an inch thick. Indium electroplated into the depressions formed the collector and emitter.

The first "prototype" pocket transistor radio was shown by INTERMETALL (a company founded by Herbert Mataré in 1952) at the *Internationale Funkausstellung Düsseldorf* between August 29, 1953 and September 9, 1953.

The first "production" all-transistor radio was the Regency TR-1, released in October 1954. Produced as a joint venture between the Regency Division of Industrial Development Engineering Associates, I.D.E.A. and Texas Instruments of Dallas Texas, the TR-1 was manufactured in Indianapolis, Indiana. It was a near pocket-sized radio featuring 4 transistors and one germanium diode. The industrial design was outsourced to the Chicago firm of industrial design firm of Painter, Teague and Petertil. It was initially released in one of four different colours: black, bone white, red, and gray. Other colours were to shortly follow.

The first working silicon transistor was developed at Bell Labs on January 26, 1954 by Morris Tanenbaum. The first commercial silicon transistor was produced by Texas Instruments in 1954. This was the work of Gordon Teal, an expert in growing crystals of high purity, who had previously worked at Bell Labs. The first MOSFET actually built was by Kahng and Atalla at Bell Labs in 1960.

## Importance

A Darlington transistor opened up so the actual transistor chip (the small square) can be seen inside. A Darlington transistor is effectively two transistors on the same chip. One transistor is much larger than the other, but both are large in comparison to transistors in large-scale integration because this particular example is intended for power applications

The transistor is the key active component in practically all modern electronics. Many consider it to be one of the greatest inventions of the 20th century. Its importance in today's society rests on its ability to be mass-produced using a highly automated process (semiconductor device fabrication) that achieves astonishingly low per-transistor costs. The invention of the first transistor at Bell Labs was named an IEEE Milestone in 2009.

Although several companies each produce over a billion individually packaged (known as *discrete*) transistors every year, the vast majority of transistors are now produced in integrated circuits (often shortened to *IC*, *microchips* or simply *chips*), along with diodes, resistors, capacitors and other electronic components, to produce complete electronic circuits. A logic gate consists of up to about twenty transistors whereas an advanced microprocessor, as of 2009, can use as many as 3 billion transistors (MOSFETs). "About 60 million transistors were built in 2002... for [each] man, woman, and child on Earth."

The transistor's low cost, flexibility, and reliability have made it a ubiquitous device. Transistorized mechatronic circuits have replaced electromechanical devices in controlling appliances and machinery. It is often easier and cheaper to use a standard microcontroller and write a computer program to carry out a control function than to design an equivalent mechanical system to control that same function.

## Simplified Operation

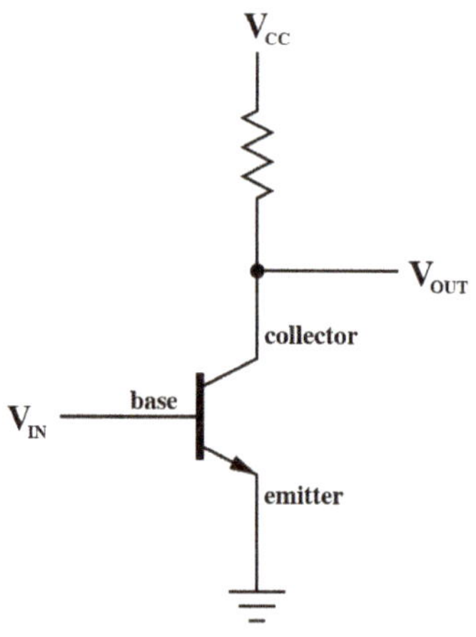


A simple circuit diagram to show the labels of a n–p–n bipolar transistor

The essential usefulness of a transistor comes from its ability to use a small signal applied between one pair of its terminals to control a much larger signal at another pair of terminals. This property is called gain. It can produce a stronger output signal, a voltage or current, which is proportional to a weaker input signal; that is, it can act as an amplifier. Alternatively, the transistor can be used to turn current on or off in a circuit as an electrically controlled switch, where the amount of current is determined by other circuit elements.

There are two types of transistors, which have slight differences in how they are used in a circuit. A *bipolar transistor* has terminals labeled base, collector, and emitter. A small current at the base terminal (that is, flowing between the base and the emitter) can control or switch a much larger current between the collector and emitter terminals. For a field-effect transistor, the terminals are labeled gate, source, and drain, and a voltage at the gate can control a current between source and drain.

The image represents a typical bipolar transistor in a circuit. Charge will flow between emitter and collector terminals depending on the current in the base. Because internally the base and emitter connections behave like a semiconductor diode, a voltage drop develops between base and emitter while the base current exists. The amount of this voltage depends on the material the transistor is made from, and is referred to as $V_{BE}$.

## Transistor as a Switch

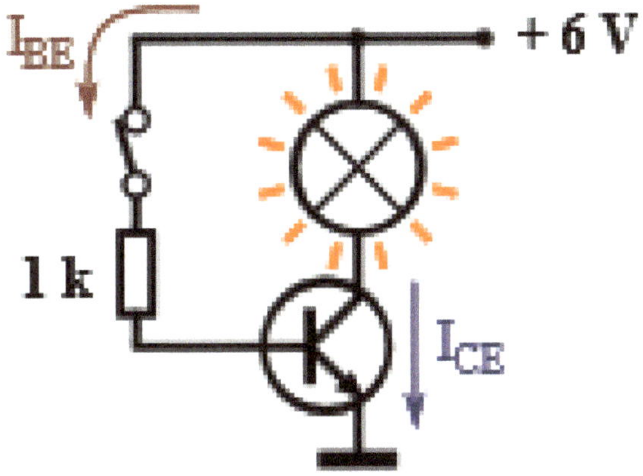


BJT used as an electronic switch, in grounded-emitter configuration

Transistors are commonly used in digital circuits as electronic switches which can be either in an "on" or "off" state, both for high-power applications such as switched-mode power supplies and for low-power applications such as logic gates. Important parameters for this application include the current switched, the voltage handled, and the switching speed, characterised by the rise and fall times.

In a grounded-emitter transistor circuit, such as the light-switch circuit shown, as the base voltage rises, the emitter and collector currents rise exponentially. The collector voltage drops because of reduced resistance from collector to emitter. If the voltage difference between the collector and emitter were zero (or near zero), the collector current would be limited only by the load resistance (light bulb) and the supply voltage. This is called *saturation* because current is flowing from collector to emitter freely. When saturated, the switch is said to be *on*.

Providing sufficient base drive current is a key problem in the use of bipolar transistors as switches. The transistor provides current gain, allowing a relatively large current in the collector to be switched by a much smaller current into the base terminal. The ratio of these currents varies depending on the type of transistor, and even for a particular type, varies depending on the collector current. In the example light-switch circuit shown, the resistor is chosen to provide enough base current to ensure the transistor will be saturated.

In a switching circuit, the idea is to simulate, as near as possible, the ideal switch having the properties of open circuit when off, short circuit when on, and an instantaneous transition between the two states. Parameters are chosen such that the "off" output is limited to leakage currents too small to affect connected circuitry; the resistance of the transistor in the "on" state is too small to affect circuitry; and the transition between the two states is fast enough not to have a detrimental effect.

## Transistor as an Amplifier

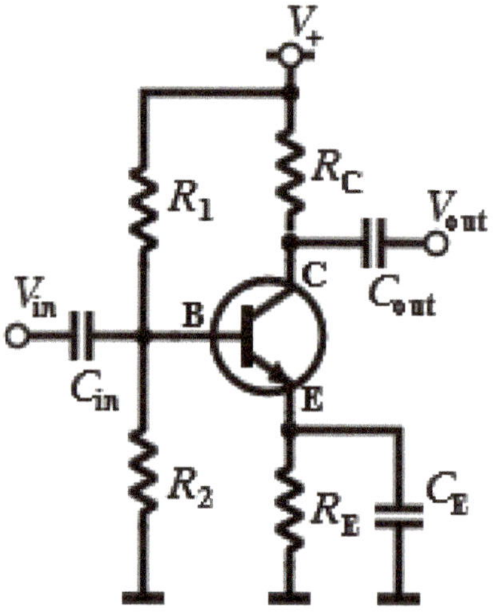


Amplifier circuit, common-emitter configuration with a voltage-divider bias circuit

The common-emitter amplifier is designed so that a small change in voltage ($V_{in}$) changes the small current through the base of the transistor; the transistor's current amplification combined with the properties of the circuit means that small swings in $V_{in}$ produce large changes in $V_{out}$.

Various configurations of single transistor amplifier are possible, with some providing current gain, some voltage gain, and some both.

From mobile phones to televisions, vast numbers of products include amplifiers for sound reproduction, radio transmission, and signal processing. The first discrete-transistor audio amplifiers barely supplied a few hundred milliwatts, but power and audio fidelity gradually increased as better transistors became available and amplifier architecture evolved.

Modern transistor audio amplifiers of up to a few hundred watts are common and relatively inexpensive.

## Comparison with Vacuum Tubes

Before transistors were developed, vacuum (electron) tubes (or in the UK "thermionic valves" or just "valves") were the main active components in electronic equipment.

## Advantages

The key advantages that have allowed transistors to replace vacuum tubes in most applications are

- no cathode heater (which produces the characteristic orange glow of tubes), reducing power consumption, eliminating delay as tube heaters warm up, and immune from cathode poisoning and depletion;
- very small size and weight, reducing equipment size;
- large numbers of extremely small transistors can be manufactured as a single integrated circuit;

- low operating voltages compatible with batteries of only a few cells;
- circuits with greater energy efficiency are usually possible. For low-power applications (e.g., voltage amplification) in particular, energy consumption can be very much less than for tubes;
- inherent reliability and very long life; tubes always degrade and fail over time. Some transistorized devices have been in service for more than 50 years ;
- complementary devices available, providing design flexibility including complementary-symmetry circuits, not possible with vacuum tubes;
- very low sensitivity to mechanical shock and vibration, providing physical ruggedness and virtually eliminating shock-induced spurious signals (e.g., microphonics in audio applications);
- not susceptible to breakage of a glass envelope, leakage, outgassing, and other physical damage.

## Limitations

Transistors have the following limitations:

- silicon transistors can age and fail;
- high-power, high-frequency operation, such as that used in over-the-air television broadcasting, is better achieved in vacuum tubes due to improved electron mobility in a vacuum;
- solid-state devices are susceptible to damage from very brief electrical and thermal events, including electrostatic discharge in handling; vacuum tubes are electrically much more rugged;
- sensitivity to radiation and cosmic rays (special radiation-hardened chips are used for spacecraft devices);
- vacuum tubes in audio applications create significant lower-harmonic distortion, the so-called tube sound, which some people prefer.

## Types

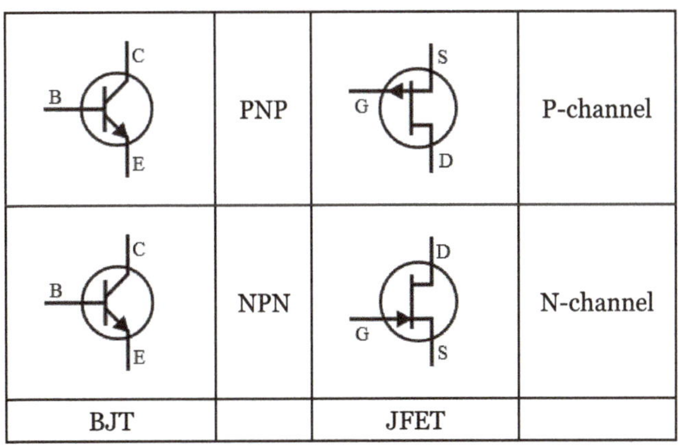


BJT and JFET symbols

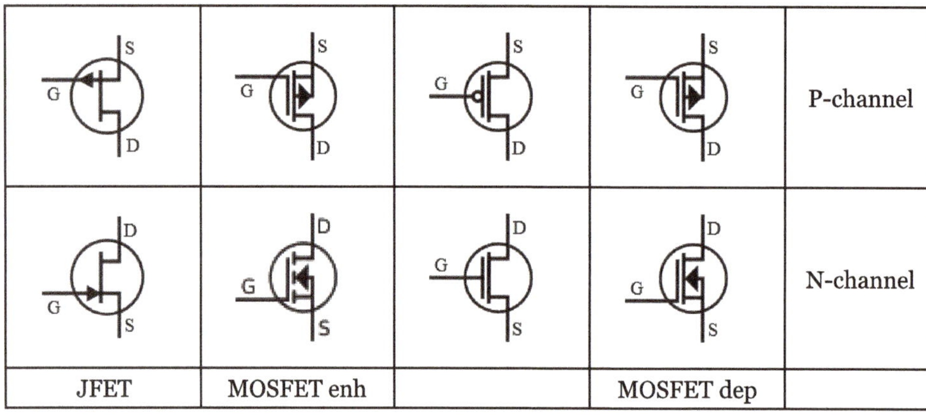


JFET and MOSFET symbols

Transistors are categorized by

- semiconductor material: the metalloids germanium (first used in 1947) and silicon (first used in 1954)—in amorphous, polycrystalline and monocrystalline form—, the compounds gallium arsenide (1966) and silicon carbide (1997), the alloy silicon-germanium (1989), the allotrope of carbon graphene (research ongoing since 2004), etc.
- structure: BJT, JFET, IGFET (MOSFET), insulated-gate bipolar transistor, "other types";
- electrical polarity (positive and negative): n–p–n, p–n–p (BJTs), n-channel, p-channel (FETs);
- maximum power rating: low, medium, high;
- maximum operating frequency: low, medium, high, radio (RF), microwave frequency (the maximum effective frequency of a transistor in a common-emitter or common-source circuit is denoted by the term $f_T$, an abbreviation for transition frequency—the frequency of transition is the frequency at which the transistor yields unity voltage gain)
- application: switch, general purpose, audio, high voltage, super-beta, matched pair;
- physical packaging: through-hole metal, through-hole plastic, surface mount, ball grid array, power modules.
- amplification factor $h_{FE}$, $\beta_F$ (transistor beta) or $g_m$ (transconductance).

Hence, a particular transistor may be described as *silicon, surface-mount, BJT, n–p–n, low-power, high-frequency switch.*

A popular way to remember which symbol represents which type of transistor is to look at the arrow and how it is arranged. Within an NPN transistor symbol, the arrow will Not Point iN. Conversely, within the PNP symbol you see that the arrow Points iN Proudly.

## Bipolar Junction Transistor (BJT)

Bipolar transistors are so named because they conduct by using both majority and minority carriers. The bipolar junction transistor, the first type of transistor to be mass-produced, is a combination of two junction diodes, and is formed of either a thin layer of p-type semiconductor sandwiched between two n-type semiconductors (an n–p–n transistor), or a thin layer of n-type semiconductor sandwiched between two p-type semiconductors (a p–n–p transistor). This construction produces two p–n junctions: a base–emitter junction and a base–collector junction, separated by a thin region of semiconductor known as the base region (two junction diodes wired together without sharing an intervening semiconducting region will not make a transistor).

BJTs have three terminals, corresponding to the three layers of semiconductor—an *emitter*, a *base*, and a *collector*. They are useful in amplifiers because the currents at the emitter and collector are controllable by a relatively small base current. In an n–p–n transistor operating in the active region, the emitter–base junction is forward biased (electrons and holes recombine at the junction), and electrons are injected into the base region. Because the base is narrow, most of these electrons will diffuse into the reverse-biased (electrons and holes are formed at, and move away from the junction) base–collector junction and be swept into the collector; perhaps one-hundredth of the electrons will recombine in the base, which is the dominant mechanism in the base current. By controlling the number of electrons that can leave the base, the number of electrons entering the collector can be controlled. Collector current is approximately β (common-emitter current gain) times the base current. It is typically greater than 100 for small-signal transistors but can be smaller in transistors designed for high-power applications.

Unlike the field-effect transistor, the BJT is a low-input-impedance device. Also, as the base–emitter voltage ($V_{BE}$) is increased the base–emitter current and hence the collector–emitter current ($I_{CE}$) increase exponentially according to the Shockley diode model and the Ebers-Moll model. Because of this exponential relationship, the BJT has a higher transconductance than the FET.

Bipolar transistors can be made to conduct by exposure to light, because absorption of photons in the base region generates a photocurrent that acts as a base current; the collector current is approximately β times the photocurrent. Devices designed for this purpose have a transparent window in the package and are called phototransistors.

## Field-effect Transistor (FET)

The *field-effect transistor*, sometimes called a *unipolar transistor*, uses either electrons (in *n-channel FET*) or holes (in *p-channel FET*) for conduction. The four terminals of the FET are named *source*, *gate*, *drain*, and *body* (*substrate*). On most FETs, the body is connected to the source inside the package, and this will be assumed for the following description.

In a FET, the drain-to-source current flows via a conducting channel that connects the *source* region to the *drain* region. The conductivity is varied by the electric field that is produced when a voltage is applied between the gate and source terminals; hence the current flowing between the drain and source is controlled by the voltage applied between the gate and source. As the gate–source voltage ($V_{GS}$) is increased, the drain–source current ($I_{DS}$) increases exponentially for $V_{GS}$ below threshold, and then at a roughly quadratic rate ($I_{GS} \propto (V_{GS} - V_T)^2$) (where $V_T$ is the threshold voltage at which drain current begins) in the "space-charge-limited" region above threshold. A quadratic behavior is not observed in modern devices, for example, at the 65 nm technology node.

For low noise at narrow bandwidth the higher input resistance of the FET is advantageous.

FETs are divided into two families: *junction FET* (JFET) and *insulated gate FET* (IGFET). The IGFET is more commonly known as a *metal–oxide–semiconductor FET* (MOSFET), reflecting its original construction from layers of metal (the gate), oxide (the insulation), and semiconductor. Unlike IGFETs, the JFET gate forms a p–n diode with the channel which lies between the source and drain. Functionally, this makes the n-channel JFET the solid-state equivalent of the vacuum tube triode which, similarly, forms a diode between its grid and cathode. Also, both devices operate in the *depletion mode*, they both have a high input impedance, and they both conduct current under the control of an input voltage.

Metal–semiconductor FETs (MESFETs) are JFETs in which the reverse biased p–n junction is replaced by a metal–semiconductor junction. These, and the HEMTs (high-electron-mobility transistors, or HFETs), in which a two-dimensional electron gas with very high carrier mobility is used for charge transport, are especially suitable for use at very high frequencies (microwave frequencies; several GHz).

FETs are further divided into *depletion-mode* and *enhancement-mode* types, depending on whether the channel is turned on or off with zero gate-to-source voltage. For enhancement mode, the channel is off at zero bias, and a gate potential can "enhance" the conduction. For the depletion mode, the channel is on at zero bias, and a gate potential (of the opposite polarity) can "deplete" the channel, reducing conduction. For either mode, a more positive gate voltage corresponds to a higher current for n-channel devices and a lower current for p-channel devices. Nearly all JFETs are depletion-mode because the diode junctions would forward bias and conduct if they were enhancement-mode devices; most IGFETs are enhancement-mode types.

## Usage of Bipolar and Field-effect Transistors

The bipolar junction transistor (BJT) was the most commonly used transistor in the 1960s and 70s. Even after MOSFETs became widely available, the BJT remained the transistor of choice for many analog circuits such as amplifiers because of their greater linearity and ease of manufacture. In integrated circuits, the desirable properties of MOSFETs allowed them to capture nearly all market share for digital circuits. Discrete MOSFETs can be applied in transistor applications, including analog circuits, voltage regulators, amplifiers, power transmitters and motor drivers.

## Other Transistor Types

Transistor symbol created on Portuguese pavement in the University of Aveiro

*For early bipolar transistors, see Bipolar junction transistor § Bipolar transistors.*

- Bipolar junction transistor (BJT):
  - heterojunction bipolar transistor, up to several hundred GHz, common in modern ultrafast and RF circuits;
  - Schottky transistor;
  - avalanche transistor:
  - Darlington transistors are two BJTs connected together to provide a high current gain equal to the product of the current gains of the two transistors;
  - insulated-gate bipolar transistors (IGBTs) use a medium-power IGFET, similarly connected to a power BJT, to give a high input impedance. Power diodes are often connected between certain terminals depending on specific use. IGBTs are particularly suitable for heavy-duty industrial applications. The ASEA Brown Boveri (ABB) *5SNA2400E170100* illustrates just how far power semiconductor technology has advanced. Intended for three-phase power supplies, this device houses three n–p–n IGBTs in a case measuring 38 by 140 by 190 mm and weighing 1.5 kg. Each IGBT is rated at 1,700 volts and can handle 2,400 amperes;
  - phototransistor;
  - multiple-emitter transistor, used in transistor–transistor logic and integrated current mirrors;
  - multiple-base transistor, used to amplify very-low-level signals in noisy environments such as the pickup of a record player or radio front ends. Effectively, it is a very large number of transistors in parallel where, at the output, the signal is added constructively, but random noise is added only stochastically.
- Field-effect transistor (FET):
  - carbon nanotube field-effect transistor (CNFET), where the channel material is replaced by a carbon nanotube;
  - junction gate field-effect transistor (JFET), where the gate is insulated by a reverse-biased p–n junction;
  - metal–semiconductor field-effect transistor (MESFET), similar to JFET with a Schottky junction instead of a p–n junction;
    - high-electron-mobility transistor (HEMT);
  - metal–oxide–semiconductor field-effect transistor (MOSFET), where the gate is insulated by a shallow layer of insulator;
  - inverted-T field-effect transistor (ITFET);
  - fin field-effect transistor (FinFET), source/drain region shapes fins on the silicon surface;

  - fast-reverse epitaxial diode field-effect transistor (FREDFET);
  - thin-film transistor, in LCDs;
  - organic field-effect transistor (OFET), in which the semiconductor is an organic compound;
  - ballistic transistor;
  - floating-gate transistor, for non-volatile storage;
  - FETs used to sense environment;
    - ion-sensitive field-effect transistor (IFSET), to measure ion concentrations in solution,
    - electrolyte–oxide–semiconductor field-effect transistor (EOSFET), neurochip,
    - deoxyribonucleic acid field-effect transistor (DNAFET).
- Tunnel field-effect transistor, where it switches by modulating quantum tunnelling through a barrier.
- Diffusion transistor, formed by diffusing dopants into semiconductor substrate; can be both BJT and FET.
- Unijunction transistor, can be used as simple pulse generators. It comprise a main body of either P-type or N-type semiconductor with ohmic contacts at each end (terminals *Base1* and *Base2*). A junction with the opposite semiconductor type is formed at a point along the length of the body for the third terminal (*Emitter*).
- Single-electron transistors (SET), consist of a gate island between two tunneling junctions. The tunneling current is controlled by a voltage applied to the gate through a capacitor.
- Nanofluidic transistor, controls the movement of ions through sub-microscopic, water-filled channels.
- Multigate devices:
  - tetrode transistor;
  - pentode transistor;
  - trigate transistor (prototype by Intel);
  - dual-gate field-effect transistors have a single channel with two gates in cascode; a configuration optimized for *high-frequency amplifiers*, *mixers*, and oscillators.
- Junctionless nanowire transistor (JNT), uses a simple nanowire of silicon surrounded by an electrically isolated "wedding ring" that acts to gate the flow of electrons through the wire.

- Vacuum-channel transistor, when in 2012, NASA and the National Nanofab Center in South Korea were reported to have built a prototype vacuum-channel transistor in only 150 nanometers in size, can be manufactured cheaply using standard silicon semiconductor processing, can operate at high speeds even in hostile environments, and could consume just as much power as a standard transistor.
- Organic electrochemical transistor.

## Part numbering Standards/Specifications

The types of some transistors can be parsed from the part number. There are three major semiconductor naming standards; in each the alphanumeric prefix provides clues to type of the device.

## Japanese Industrial Standard (JIS)

| JIS Transistor Prefix Table | |
|---|---|
| **Prefix** | **Type of transistor** |
| 2SA | high-frequency p–n–p BJTs |
| 2SB | audio-frequency p–n–p BJTs |
| 2SC | high-frequency n–p–n BJTs |
| 2SD | audio-frequency n–p–n BJTs |
| 2SJ | P-channel FETs (both JFETs and MOSFETs) |
| 2SK | N-channel FETs (both JFETs and MOSFETs) |

The *JIS-C-7012* specification for transistor part numbers starts with "2S", e.g. 2SD965, but sometimes the "2S" prefix is not marked on the package – a 2SD965 might only be marked "D965"; a 2SC1815 might be listed by a supplier as simply "C1815". This series sometimes has suffixes (such as "R", "O", "BL", standing for "red", "orange", "blue", etc.) to denote variants, such as tighter $h_{FE}$ (gain) groupings.

## European Electronic Component Manufacturers Association (EECA)

The Pro Electron standard, the European Electronic Component Manufacturers Association part numbering scheme, begins with two letters: the first gives the semiconductor type (A for germanium, B for silicon, and C for materials like GaAs); the second letter denotes the intended use (A for diode, C for general-purpose transistor, etc.). A 3-digit sequence number (or one letter then 2 digits, for industrial types) follows. With early devices this indicated the case type. Suffixes may be used, with a letter (e.g. "C" often means high $h_{FE}$, such as in: BC549C) or other codes may follow to show gain (e.g. BC327-25) or voltage rating (e.g. BUK854-800A). The more common prefixes are:

## Joint Electron Device Engineering Council (JEDEC)

The JEDEC *EIA370* transistor device numbers usually start with "2N", indicating a three-terminal device (dual-gate field-effect transistors are four-terminal devices, so begin with 3N), then a 2, 3 or 4-digit sequential number with no significance as to device properties (although early devices with low numbers tend to be germanium). For example, 2N3055 is a silicon n–p–n power transistor, 2N1301 is a p–n–p germanium switching transistor. A letter suffix (such as "A") is sometimes used to indicate a newer variant, but rarely gain groupings.

## Proprietary

Manufacturers of devices may have their own proprietary numbering system, for example CK722. Since devices are second-sourced, a manufacturer's prefix (like "MPF" in MPF102, which originally would denote a Motorola FET) now is an unreliable indicator of who made the device. Some proprietary naming schemes adopt parts of other naming schemes, for example a PN2222A is a (possibly Fairchild Semiconductor) 2N2222A in a plastic case (but a PN108 is a plastic version of a BC108, not a 2N108, while the PN100 is unrelated to other xx100 devices).

Military part numbers sometimes are assigned their own codes, such as the British Military CV Naming System.

Manufacturers buying large numbers of similar parts may have them supplied with "house numbers", identifying a particular purchasing specification and not necessarily a device with a standardized registered number. For example, an HP part 1854,0053 is a (JEDEC) 2N2218 transistor which is also assigned the CV number: CV7763

## Naming Problems

With so many independent naming schemes, and the abbreviation of part numbers when printed on the devices, ambiguity sometimes occurs. For example, two different devices may be marked "J176" (one the J176 low-power JFET, the other the higher-powered MOSFET 2SJ176).

As older "through-hole" transistors are given surface-mount packaged counterparts, they tend to be assigned many different part numbers because manufacturers have their own systems to cope with the variety in pinout arrangements and options for dual or matched n–p–n+p–n–p devices in one pack. So even when the original device (such as a 2N3904) may have been assigned by a standards authority, and well known by engineers over the years, the new versions are far from standardized in their naming.

## Construction

## Semiconductor Material

| Semiconductor material characteristics | | | | |
|---|---|---|---|---|
| **Semiconductor material** | **Junction forward voltage V @ 25 °C** | **Electron mobility $m^2/(V \cdot s)$ @ 25 °C** | **Hole mobility $m^2/(V \cdot s)$ @ 25 °C** | **Max. junction temp. °C** |
| **Ge** | 0.27 | 0.39 | 0.19 | 70 to 100 |
| **Si** | 0.71 | 0.14 | 0.05 | 150 to 200 |
| **GaAs** | 1.03 | 0.85 | 0.05 | 150 to 200 |
| **Al-Si junction** | 0.3 | — | — | 150 to 200 |

The first BJTs were made from germanium (Ge). Silicon (Si) types currently predominate but certain advanced microwave and high-performance versions now employ the *compound semiconductor* material gallium arsenide (GaAs) and the *semiconductor alloy* silicon germanium (SiGe). Single element semiconductor material (Ge and Si) is described as *elemental*.

Rough parameters for the most common semiconductor materials used to make transistors are given in the adjacent table; these parameters will vary with increase in temperature, electric field, impurity level, strain, and sundry other factors.

The *junction forward voltage* is the voltage applied to the emitter–base junction of a BJT in order to make the base conduct a specified current. The current increases exponentially as the junction forward voltage is increased. The values given in the table are typical for a current of 1 mA (the same values apply to semiconductor diodes). The lower the junction forward voltage the better, as this means that less power is required to "drive" the transistor. The junction forward voltage for a given current decreases with increase in temperature. For a typical silicon junction the change is −2.1 mV/°C. In some circuits special compensating elements (sensistors) must be used to compensate for such changes.

The density of mobile carriers in the channel of a MOSFET is a function of the electric field forming the channel and of various other phenomena such as the impurity level in the channel. Some impurities, called dopants, are introduced deliberately in making a MOSFET, to control the MOSFET electrical behavior.

The *electron mobility* and *hole mobility* columns show the average speed that electrons and holes diffuse through the semiconductor material with an electric field of 1 volt per meter applied across the material. In general, the higher the electron mobility the faster the transistor can operate. The table indicates that Ge is a better material than Si in this respect. However, Ge has four major shortcomings compared to silicon and gallium arsenide:

- Its maximum temperature is limited;
- it has relatively high leakage current;
- it cannot withstand high voltages;
- it is less suitable for fabricating integrated circuits.

Because the electron mobility is higher than the hole mobility for all semiconductor materials, a given bipolar n–p–n transistor tends to be swifter than an equivalent p–n–p transistor. GaAs has the highest electron mobility of the three semiconductors. It is for this reason that GaAs is used in high-frequency applications. A relatively recent FET development, the *high-electron-mobility transistor* (HEMT), has a heterostructure (junction between different semiconductor materials) of aluminium gallium arsenide (AlGaAs)-gallium arsenide (GaAs) which has twice the electron mobility of a GaAs-metal barrier junction. Because of their high speed and low noise, HEMTs are used in satellite receivers working at frequencies around 12 GHz. HEMTs based on gallium nitride and aluminium gallium nitride (AlGaN/GaN HEMTs) provide a still higher electron mobility and are being developed for various applications.

Max. junction temperature values represent a cross section taken from various manufacturers' data sheets. This temperature should not be exceeded or the transistor may be damaged.

Al–Si junction refers to the high-speed (aluminum–silicon) metal–semiconductor barrier diode, commonly known as a Schottky diode. This is included in the table because some silicon power IGFETs have a *parasitic* reverse Schottky diode formed between the source and drain as part of the fabrication process. This diode can be a nuisance, but sometimes it is used in the circuit.

## Packaging

Assorted discrete transistors

Discrete transistors are individually packaged transistors. Transistors come in many different semiconductor packages. The two main categories are through-hole (or leaded), and surface-mount, also known as surface-mount device (SMD). The ball grid array (BGA) is the latest surface-mount package (currently only for large integrated circuits). It has solder "balls" on the underside in place of leads. Because they are smaller and have shorter interconnections, SMDs have better high-frequency characteristics but lower power rating.

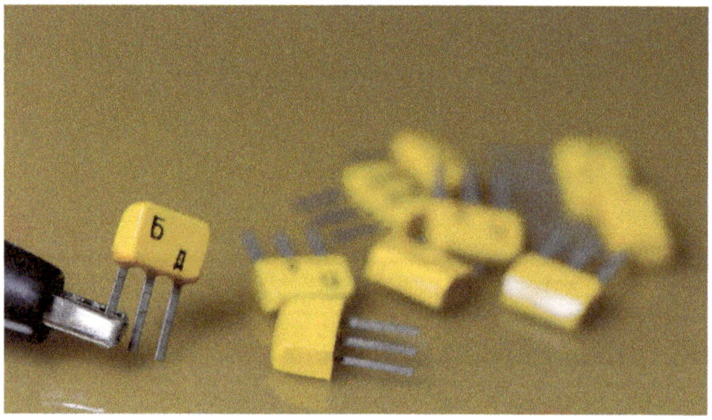

Soviet KT315b transistors

Transistor packages are made of glass, metal, ceramic, or plastic. The package often dictates the power rating and frequency characteristics. Power transistors have larger packages that can be clamped to heat sinks for enhanced cooling. Additionally, most power transistors have the collector or drain physically connected to the metal enclosure. At the other extreme, some surface-mount *microwave* transistors are as small as grains of sand.

Often a given transistor type is available in several packages. Transistor packages are mainly standardized, but the assignment of a transistor's functions to the terminals is not: other transistor types can assign other functions to the package's terminals. Even for the same transistor type the terminal assignment can vary (normally indicated by a suffix letter to the part number, q.e. BC212L and BC212K).

Nowadays most transistors come in a wide range of SMT packages, in comparison the list of available through-hole packages is relatively small, here is a short list of the most common through-hole transistors packages in alphabetical order: ATV, E-line, MRT, HRT, SC-43, SC-72, TO-3, TO-18, TO-39, TO-92, TO-126, TO220, TO247, TO251, TO262, ZTX851

## Flexible Transistors

Researchers have made several kinds of flexible transistors, including organic field-effect transistors. Flexible transistors are useful in some kinds of flexible displays and other flexible electronics.

## Light-emitting Diode

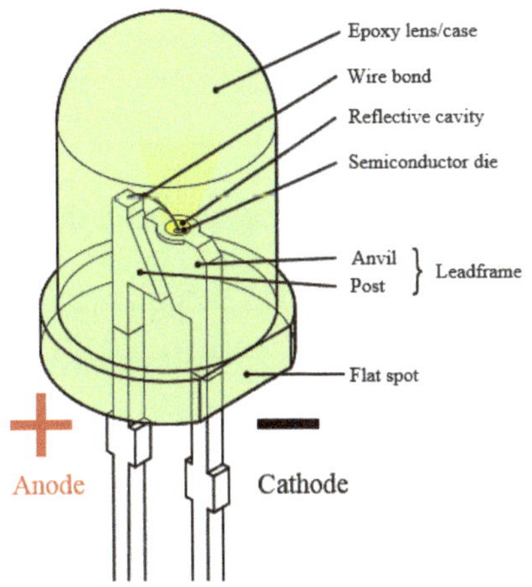


Parts of a conventional LED. The flat bottom surfaces of the anvil and post embedded inside the epoxy act as anchors, to prevent the conductors from being forcefully pulled out via mechanical strain or vibration

A bulb-shaped modern retrofit LED lamp with aluminium heat sink, a light diffusing dome and E27 screw base, using a built-in power supply working on mains voltage

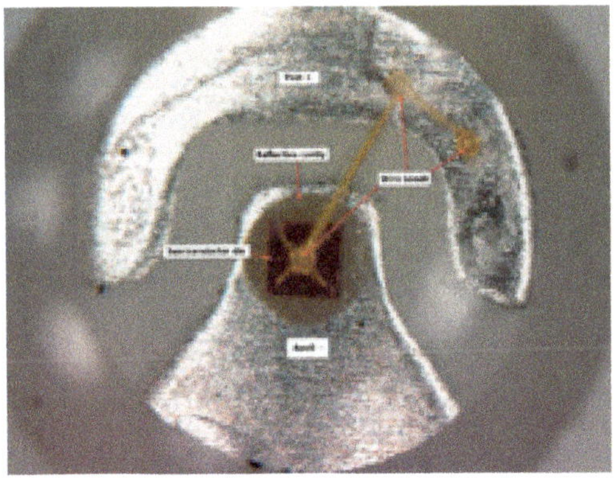

Close up image of a surface mount LED

A light-emitting diode (LED) is a two-lead semiconductor light source. It is a p–n junction diode that emits light when activated. When a suitable voltage is applied to the leads, electrons are able to recombine with electron holes within the device, releasing energy in the form of photons. This effect is called electroluminescence, and the color of the light (corresponding to the energy of the photon) is determined by the energy band gap of the semiconductor. LEDs are typically small (less than 1 $mm^2$ ) and integrated optical components may be used to shape the radiation pattern.

Appearing as practical electronic components in 1962, the earliest LEDs emitted low-intensity infrared light. Infrared LEDs are still frequently used as transmitting elements in remote-control circuits, such as those in remote controls for a wide variety of consumer electronics. The first visible-light LEDs were also of low intensity and limited to red. Modern LEDs are available across the visible, ultraviolet, and infrared wavelengths, with very high brightness.

Early LEDs were often used as indicator lamps for electronic devices, replacing small incandescent bulbs. They were soon packaged into numeric readouts in the form of seven-segment displays and were commonly seen in digital clocks. Recent developments have produced LEDs suitable for environmental and task lighting. LEDs have led to new displays and sensors, while their high switching rates are useful in advanced communications technology.

LEDs have many advantages over incandescent light sources, including lower energy consumption, longer lifetime, improved physical robustness, smaller size, and faster switching. Light-emitting diodes are used in applications as diverse as aviation lighting, automotive headlamps, advertising, general lighting, traffic signals, camera flashes, and lighted wallpaper. As of 2017, LED lights home room lighting are as cheap or cheaper than compact fluorescent lamp sources of comparable output. They are also significantly more energy efficient and, arguably, have fewer environmental concerns linked to their disposal.

## History

### Discoveries and Early Devices

Electroluminescence as a phenomenon was discovered in 1907 by the British experimenter H. J. Round of Marconi Labs, using a crystal of silicon carbide and a cat's-whisker detector. Russian inventor Oleg Losev reported creation of the first LED in 1927. His research was distributed in Soviet, German and British scientific journals, but no practical use was made of the discovery for several decades. Kurt Lehovec, Carl Accardo, and Edward Jamgochian explained these first light-emitting diodes in 1951 using an apparatus employing SiC crystals with a current source of battery or pulse generator and with a comparison to a variant, pure, crystal in 1953.

Green electroluminescence from a point contact on a crystal of SiC recreates Round's original experiment from 1907

Rubin Braunstein of the Radio Corporation of America reported on infrared emission from gallium arsenide (GaAs) and other semiconductor alloys in 1955. Braunstein observed infrared emission generated by simple diode structures using gallium antimonide (GaSb), GaAs, indium phosphide (InP), and silicon-germanium (SiGe) alloys at room temperature and at 77 Kelvin.

In 1957, Braunstein further demonstrated that the rudimentary devices could be used for non-radio communication across a short distance. As noted by Kroemer Braunstein "...had set up a simple optical communications link: Music emerging from a record player was used via suitable electronics to modulate the forward current of a GaAs diode. The emitted light was detected by a PbS diode some distance away. This signal was fed into an audio amplifier and played back by a loudspeaker. Intercepting the beam stopped the music. We had a great deal of fun playing with this setup." This setup presaged the use of LEDs for optical communication applications.

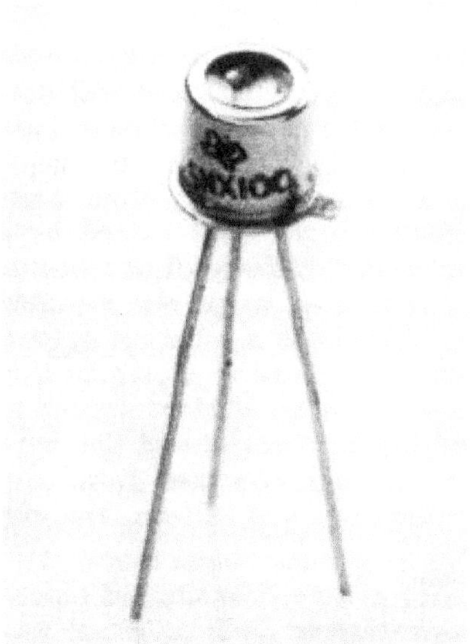

A Texas Instruments SNX-100 GaAs LED contained in a TO-18 transistor metal case

In September 1961, while working at Texas Instruments in Dallas, Texas, James R. Biard and Gary Pittman discovered near-infrared (900 nm) light emission from a tunnel diode they had constructed on a GaAs substrate. By October 1961, they had demonstrated efficient light emission and signal coupling between a GaAs p-n junction light emitter and an electrically-isolated semiconductor photodetector. On August 8, 1962, Biard and Pittman filed a patent titled "Semiconductor Radiant Diode" based on their findings, which described a zinc diffused p–n junction LED with a spaced cathode contact to allow for efficient emission of infrared light under forward bias. After establishing the priority of their work based on engineering notebooks predating submissions from G.E. Labs, RCA Research Labs, IBM Research Labs, Bell Labs, and Lincoln Lab at MIT, the U.S. patent office issued the two inventors the patent for the GaAs infrared (IR) light-emitting diode (U.S. Patent US3293513), the first practical LED. Immediately after filing the patent, Texas Instruments (TI) began a project to manufacture infrared diodes. In October 1962, TI announced the first commercial LED product (the SNX-100), which employed a pure GaAs crystal to emit a 890 nm light output. In October 1963, TI announced the first commercial hemispherical LED, the SNX-110.

The first visible-spectrum (red) LED was developed in 1962 by Nick Holonyak, Jr. while working at General Electric. Holonyak first reported his LED in the journal *Applied Physics Letters* on

December 1, 1962. M. George Craford, a former graduate student of Holonyak, invented the first yellow LED and improved the brightness of red and red-orange LEDs by a factor of ten in 1972. In 1976, T. P. Pearsall created the first high-brightness, high-efficiency LEDs for optical fiber telecommunications by inventing new semiconductor materials specifically adapted to optical fiber transmission wavelengths.

## Initial Commercial Development

The first commercial LEDs were commonly used as replacements for incandescent and neon indicator lamps, and in seven-segment displays, first in expensive equipment such as laboratory and electronics test equipment, then later in such appliances as TVs, radios, telephones, calculators, as well as watches. Until 1968, visible and infrared LEDs were extremely costly, in the order of US$200 per unit, and so had little practical use. The Monsanto Company was the first organization to mass-produce visible LEDs, using gallium arsenide phosphide (GaAsP) in 1968 to produce red LEDs suitable for indicators. Hewlett Packard (HP) introduced LEDs in 1968, initially using GaAsP supplied by Monsanto. These red LEDs were bright enough only for use as indicators, as the light output was not enough to illuminate an area. Readouts in calculators were so small that plastic lenses were built over each digit to make them legible. Later, other colors became widely available and appeared in appliances and equipment. In the 1970s commercially successful LED devices at less than five cents each were produced by Fairchild Optoelectronics. These devices employed compound semiconductor chips fabricated with the planar process invented by Dr. Jean Hoerni at Fairchild Semiconductor. The combination of planar processing for chip fabrication and innovative packaging methods enabled the team at Fairchild led by optoelectronics pioneer Thomas Brandt to achieve the needed cost reductions. LED producers continue to use these methods.

LED display of a TI-30 scientific calculator (ca. 1978), which uses plastic lenses to increase the visible digit size

Most LEDs were made in the very common 5 mm T1¾ and 3 mm T1 packages, but with rising power output, it has grown increasingly necessary to shed excess heat to maintain reliability, so more complex packages have been adapted for efficient heat dissipation. Packages for state-of-the-art high-power LEDs bear little resemblance to early LEDs.

## Blue LED

Blue LEDs were first developed by Herbert Paul Maruska at RCA in 1972 using gallium nitride (GaN) on a sapphire substrate. SiC-types were first commercially sold in the United States by Cree in 1989. However, neither of these initial blue LEDs were very bright.

The first high-brightness blue LED was demonstrated by Shuji Nakamura of Nichia Corporation in 1994 and was based on InGaN. In parallel, Isamu Akasaki and Hiroshi Amano in Nagoya were working on developing the important GaN nucleation on sapphire substrates and the demonstration of p-type doping of GaN. Nakamura, Akasaki, and Amano were awarded the 2014 Nobel prize in physics for their work. In 1995, Alberto Barbieri at the Cardiff University Laboratory (GB) investigated the efficiency and reliability of high-brightness LEDs and demonstrated a "transparent contact" LED using indium tin oxide (ITO) on (AlGaInP/GaAs).

In 2001 and 2002, processes for growing gallium nitride (GaN) LEDs on silicon were successfully demonstrated. In January 2012, Osram demonstrated high-power InGaN LEDs grown on silicon substrates commercially.

## White LEDs and the Illumination Breakthrough

The attainment of high efficiency in blue LEDs was quickly followed by the development of the first white LED. In this device a $Y_3Al_5O_{12}$:Ce (known as "YAG") phosphor coating on the emitter absorbs some of the blue emission and produces yellow light through fluorescence. The combination of that yellow with remaining blue light appears white to the eye. However, using different phosphors (fluorescent materials) it also became possible to instead produce green and red light through fluorescence. The resulting mixture of red, green and blue is not only perceived by humans as white light but is superior for illumination in terms of color rendering, whereas one cannot appreciate the color of red or green objects illuminated only by the yellow (and remaining blue) wavelengths from the YAG phosphor.

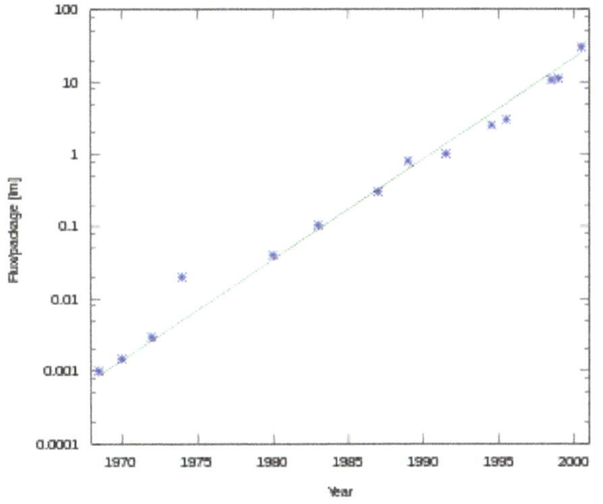


Illustration of Haitz's law, showing improvement in light output per LED over time, with a logarithmic scale on the vertical axis

The first white LEDs were expensive and inefficient. However, the light output of LEDs has increased exponentially, with a doubling occurring approximately every 36 months since the 1960s (similar to Moore's law). This trend is generally attributed to the parallel development of other semiconductor technologies and advances in optics and materials science and has been called Haitz's law after Dr. Roland Haitz.

Light output and efficiency of blue and near-ultraviolet LEDs rose as the cost of reliable devices fell. This led to relatively high-power white-light LEDs for illumination, which are replacing incandescent and fluorescent lighting.

Experimental white LEDs have been demonstrated to produce over 300 lumens per watt of electricity; some can last up to 100,000 hours. Compared to incandescent bulbs, this is not only a huge increase in electrical efficiency but – over time – a similar or lower cost per bulb.

## Working Principle

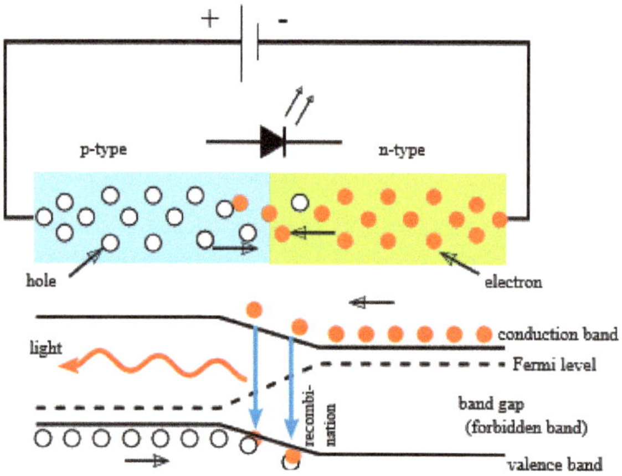


The inner workings of an LED, showing circuit (top) and band diagram (bottom)

A P-N junction can convert absorbed light energy into a proportional electric current. The same process is reversed here (i.e. the P-N junction emits light when electrical energy is applied to it). This phenomenon is generally called electroluminescence, which can be defined as the emission of light from a semi-conductor under the influence of an electric field. The charge carriers recombine in a forward-biased P-N junction as the electrons cross from the N-region and recombine with the holes existing in the P-region. Free electrons are in the conduction band of energy levels, while holes are in the valence energy band. Thus the energy level of the holes is less than the energy levels of the electrons. Some portion of the energy must be dissipated to recombine the electrons and the holes. This energy is emitted in the form of heat and light.

The electrons dissipate energy in the form of heat for silicon and germanium diodes but in gallium arsenide phosphide (GaAsP) and gallium phosphide (GaP) semiconductors, the electrons dissipate energy by emitting photons. If the semiconductor is translucent, the junction becomes the source of light as it is emitted, thus becoming a light-emitting diode. However, when the junction is reverse biased, the LED produces no light and—if the potential is great enough, the device is damaged.

## Technology

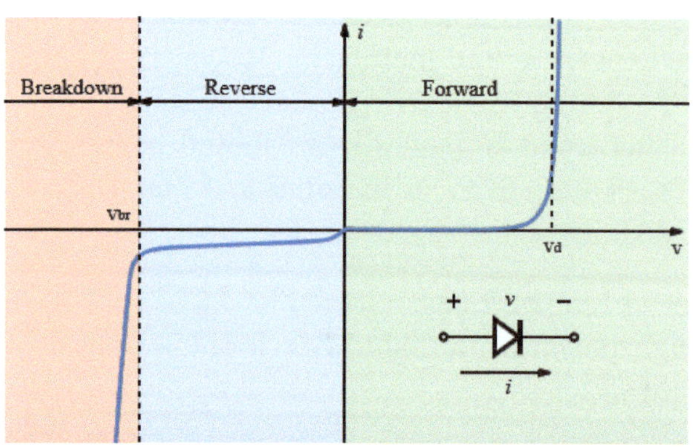


I-V diagram for a diode. An LED begins to emit light when more than 2 or 3 volts is applied. The reverse bias region uses a different vertical scale from the forward bias region to show that the leakage current is nearly constant with voltage until breakdown occurs. In forward bias, the current is small but increases exponentially with voltage

## Physics

The LED consists of a chip of semiconducting material doped with impurities to create a *p-n junction*. As in other diodes, current flows easily from the p-side, or anode, to the n-side, or cathode, but not in the reverse direction. Charge-carriers—electrons and holes—flow into the junction from electrodes with different voltages. When an electron meets a hole, it falls into a lower energy level and releases energy in the form of a photon.

The wavelength of the light emitted, and thus its color, depends on the band gap energy of the materials forming the *p-n junction*. In silicon or germanium diodes, the electrons and holes usually recombine by a *non-radiative transition*, which produces no optical emission, because these are indirect band gap materials. The materials used for the LED have a direct band gap with energies corresponding to near-infrared, visible, or near-ultraviolet light.

LED development began with infrared and red devices made with gallium arsenide. Advances in materials science have enabled making devices with ever-shorter wavelengths, emitting light in a variety of colors.

LEDs are usually built on an n-type substrate, with an electrode attached to the p-type layer deposited on its surface. P-type substrates, while less common, occur as well. Many commercial LEDs, especially GaN/InGaN, also use sapphire substrate.

## Refractive Index

Bare uncoated semiconductors such as silicon exhibit a very high refractive index relative to open air, which prevents passage of photons arriving at sharp angles relative to the air-contacting surface of the semiconductor due to total internal reflection. This property affects both the light-emission efficiency of LEDs as well as the light-absorption efficiency of photovoltaic cells. The refractive index of silicon is 3.96 (at 590 nm), while air is 1.0002926.

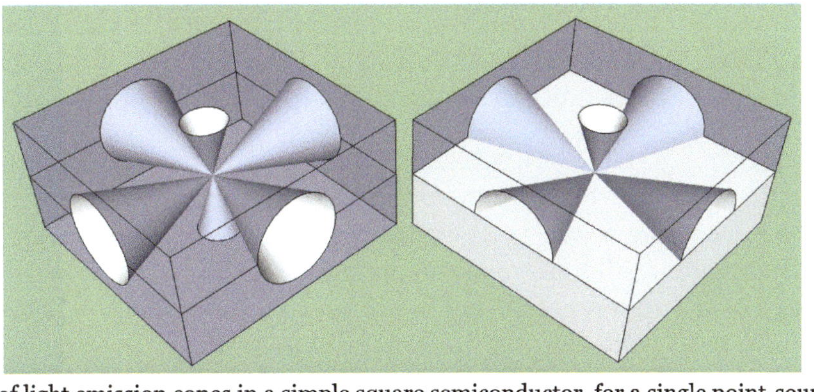

Idealized example of light emission cones in a simple square semiconductor, for a single point-source emission zone. The left illustration is for a translucent wafer, while the right illustration shows the half-cones formed when the bottom layer is opaque. The light is actually emitted equally in all directions from the point-source, but can only escape perpendicular to the semiconductor's surface and some degrees to the side, which is illustrated by the cone shapes. When the critical angle is exceeded, photons are reflected internally. The areas between the cones represent the trapped light energy wasted as heat.

Internal reflections can escape through other crystalline faces if the incidence angle is low enough and the crystal is sufficiently transparent to not re-absorb the photon emission. But for a simple square LED with 90-degree angled surfaces on all sides, the faces all act as equal angle mirrors. In this case, most of the light can not escape and is lost as waste heat in the crystal.

Most materials used for LED production have very high refractive indices. This means that much of the light is reflected back into the material at the material/air surface interface. Thus, light extraction in LEDs is an important aspect of LED production, subject to much research and development. The light emission cones of a real LED wafer are far more complex than a single point-source light emission. The light emission zone is typically a two-dimensional plane between the wafers. Every atom across this plane has an individual set of emission cones. Drawing the billions of overlapping cones is impossible, so this is a simplified diagram showing the extents of all the emission cones combined. The larger side cones are clipped to show the interior features and reduce image complexity; they would extend to the opposite edges of the two-dimensional emission plane.

In general, a flat-surface uncoated LED semiconductor chip emits light only perpendicular to the semiconductor's surface, and a few degrees to the side, in a cone shape referred to as the *light cone*, *cone of light*, or the *escape cone*. The maximum angle of incidence is referred to as the critical angle. When this angle is exceeded, photons no longer escape the semiconductor but are instead reflected internally inside the semiconductor crystal as if it were a mirror.

A convoluted chip surface with angled facets similar to a jewel or fresnel lens can increase light output by distributing light perpendicular to the chip surface and far to the sides of the photon emission point.

The ideal shape of a semiconductor with maximum light output would be a microsphere with the photon emission occurring at the exact center, with electrodes penetrating to the center to contact at the emission point. All light rays emanating from the center would be perpendicular to the entire surface of the sphere, resulting in no internal reflections. A hemispherical semiconductor would also work, with the flat back-surface serving as a mirror to back-scattered photons.

## Transition Coatings

After the doping of the wafer, it is cut apart into individual dies. Each die is commonly called a chip.

Many LED semiconductor chips are encapsulated or potted in clear or colored molded plastic shells. The plastic shell has three purposes:

1. Mounting the semiconductor chip in devices is easier to accomplish.
2. The tiny fragile electrical wiring is physically supported and protected from damage.
3. The plastic acts as a refractive intermediary between the relatively high-index semiconductor and low-index open air.

The third feature helps to boost the light emission from the semiconductor by acting as a diffusing lens, emitting light at a much higher angle of incidence from the light cone than the bare chip would alone.

## Efficiency and Operational Parameters

Typical indicator LEDs are designed to operate with no more than 30–60 milliwatts (mW) of electrical power. Around 1999, Philips Lumileds introduced power LEDs capable of continuous use at one watt. These LEDs used much larger semiconductor die sizes to handle the large power inputs. Also, the semiconductor dies were mounted onto metal slugs to allow for heat removal from the LED die.

One of the key advantages of LED-based lighting sources is high luminous efficacy. White LEDs quickly matched and overtook the efficacy of standard incandescent lighting systems. In 2002, Lumileds made five-watt LEDs available with luminous efficacy of 18–22 lumens per watt (lm/W). For comparison, a conventional incandescent light bulb of 60–100 watts emits around 15 lm/W, and standard fluorescent lights emit up to 100 lm/W.

As of 2012, Philips had achieved the following efficacies for each color. The efficiency values show the physics – light power out per electrical power in. The lumen-per-watt efficacy value includes characteristics of the human eye and is derived using the luminosity function.

| Color | Wavelength range (nm) | Typical efficiency coefficient | Typical efficacy (lm/W) |
|---|---|---|---|
| Red | $620 < \lambda < 645$ | 0.39 | 72 |
| Red-orange | $610 < \lambda < 620$ | 0.29 | 98 |
| Green | $520 < \lambda < 550$ | 0.15 | 93 |
| Cyan | $490 < \lambda < 520$ | 0.26 | 75 |
| Blue | $460 < \lambda < 490$ | 0.35 | 37 |

In September 2003, a new type of blue LED was demonstrated by Cree that consumes 24 mW at 20 milliamperes (mA). This produced a commercially packaged white light giving 65 lm/W at 20 mA, becoming the brightest white LED commercially available at the time, and more than four times as efficient as standard incandescents. In 2006, they demonstrated a prototype with a record white LED luminous efficacy of 131 lm/W at 20 mA. Nichia Corporation has developed a white LED with luminous efficacy of 150 lm/W at a forward current of 20 mA. Cree's XLamp XM-L LEDs, commercially available in 2011, produce 100 lm/W at their full power of 10 W, and up to 160 lm/W at around 2 W input power. In 2012, Cree announced a white LED giving 254 lm/W, and 303 lm/W in March 2014. Practical general lighting needs high-power LEDs, of one watt or more. Typical operating currents for such devices begin at 350 mA.

These efficiencies are for the light-emitting diode only, held at low temperature in a lab. Since LEDs installed in real fixtures operate at higher temperature and with driver losses, real-world efficiencies are much lower. United States Department of Energy (DOE) testing of commercial LED lamps designed to replace incandescent lamps or CFLs showed that average efficacy was still about 46 lm/W in 2009 (tested performance ranged from 17 lm/W to 79 lm/W).

## Efficiency Droop

Efficiency droop is the decrease in luminous efficiency of LEDs as the electric current increases above tens of milliamperes.

This effect was initially theorized to be related to elevated temperatures. Scientists proved the opposite is true: though the life of an LED would be shortened, the efficiency droop is less severe at elevated temperatures. The mechanism causing efficiency droop was identified in 2007 as Auger recombination, which was taken with mixed reaction. In 2013, a study confirmed Auger recombination as the cause of efficiency droop.

In addition to being less efficient, operating LEDs at higher electric currents creates higher heat levels, which compromise LED lifetime. Because of this increased heat at higher currents, high-brightness LEDs have an industry standard of operating at only 350 mA, which is a compromise between light output, efficiency, and longevity.

## Possible Solutions

Instead of increasing current levels, luminance is usually increased by combining multiple LEDs in one bulb. Solving the problem of efficiency droop would mean that household LED light bulbs would need fewer LEDs, which would significantly reduce costs.

Researchers at the U.S. Naval Research Laboratory have found a way to lessen the efficiency droop. They found that the droop arises from non-radiative Auger recombination of the injected carriers. They created quantum wells with a soft confinement potential to lessen the non-radiative Auger processes.

Researchers at Taiwan National Central University and Epistar Corp are developing a way to lessen the efficiency droop by using ceramic aluminium nitride (AlN) substrates, which are more thermally conductive than the commercially used sapphire. The higher thermal conductivity reduces self-heating effects.

## Lifetime and Failure

Solid-state devices such as LEDs are subject to very limited wear and tear if operated at low currents and at low temperatures. Typical lifetimes quoted are 25,000 to 100,000 hours, but heat and current settings can extend or shorten this time significantly.

The most common symptom of LED (and diode laser) failure is the gradual lowering of light output and loss of efficiency. Sudden failures, although rare, can also occur. Early red LEDs were notable for their short service life. With the development of high-power LEDs, the devices are subjected to higher junction temperatures and higher current densities than traditional devices. This causes stress on the material and may cause early light-output degradation. To quantitify useful lifetime in a standardized manner, some suggest using L70 or L50, which are runtimes (typically in thousands of hours) at which a given LED reaches 70% and 50% of initial light output, respectively.

Whereas in most previous sources of light (incandescent lamps, discharge lamps, and those that burn combustible fuel, e.g. candles and oil lamps) the light results from heat, LEDs only operate if they are kept cool enough. The manufacturer commonly specifies a maximum junction temperature of 125 or 150 °C, and lower temperatures are advisable in the interests of long life. At these temperatures, relatively little heat is lost by radiation, which means that the light beam generated by an LED is cool.

The waste heat in a high-power LED (which as of 2015 can be less than half the power that it consumes) is conveyed by conduction through the substrate and package of the LED to a heat sink, which gives up the heat to the ambient air by convection. Careful thermal design is, therefore, essential, taking into account the thermal resistances of the LED's package, the heat sink and the interface between the two. Medium-power LEDs are often designed to solder directly to a printed circuit board that contains a thermally conductive metal layer. High-power LEDs are packaged in large-area ceramic packages that attach to a metal heat sink—the interface being a material with high thermal conductivity (thermal grease, phase-change material, thermally conductive pad, or thermal adhesive).

If an LED-based lamp is installed in an unventilated luminaire, or a luminaire is located in an environment that does not have free air circulation, the LED is likely to overheat, resulting in reduced life or early catastrophic failure. Thermal design is often based on an ambient temperature of 25 °C (77 °F). LEDs used in outdoor applications, such as traffic signals or in-pavement signal lights, and in climates where the temperature within the light fixture gets very high, could experience reduced output or even failure.

Since LED efficacy is higher at low temperatures, LED technology is well suited for supermarket freezer lighting. Because LEDs produce less waste heat than incandescent lamps, their use in freezers can save on refrigeration costs as well. However, they may be more susceptible to frost and snow buildup than incandescent lamps, so some LED lighting systems have been designed with an added heating circuit. Additionally, research has developed heat sink technologies that transfer heat produced within the junction to appropriate areas of the light fixture.

## Colors and Materials

Conventional LEDs are made from a variety of inorganic semiconductor materials. The following table shows the available colors with wavelength range, voltage drop, and material:

| Color | Wavelength [nm] | Voltage drop [ΔV] | Semiconductor material |
|---|---|---|---|
| Infrared | $\lambda > 760$ | $\Delta V < 1.63$ | Gallium arsenide (GaAs) Aluminium gallium arsenide (AlGaAs) |
| Red | $610 < \lambda < 760$ | $1.63 < \Delta V < 2.03$ | Aluminium gallium arsenide (AlGaAs) Gallium arsenide phosphide (GaAsP) Aluminium gallium indium phosphide (AlGaInP) Gallium(III) phosphide (GaP) |
| Orange | $590 < \lambda < 610$ | $2.03 < \Delta V < 2.10$ | Gallium arsenide phosphide (GaAsP) Aluminium gallium indium phosphide (AlGaInP) Gallium(III) phosphide (GaP) |

| | | | |
|---|---|---|---|
| Yellow | $570 < \lambda < 590$ | $2.10 < \Delta V < 2.18$ | Gallium arsenide phosphide (GaAsP) Aluminium gallium indium phosphide (AlGaInP) Gallium(III) phosphide (GaP) |
| Green | $500 < \lambda < 570$ | $1.9 < \Delta V < 4.0$ | **Traditional green:** Gallium(III) phosphide (GaP) Aluminium gallium indium phosphide (AlGaInP) Aluminium gallium phosphide (AlGaP) **Pure green:** Indium gallium nitride (InGaN) / Gallium(III) nitride (GaN) |
| Blue | $450 < \lambda < 500$ | $2.48 < \Delta V < 3.7$ | Zinc selenide (ZnSe) Indium gallium nitride (InGaN) Silicon carbide (SiC) as substrate Silicon (Si) as substrate—under development |
| Violet | $400 < \lambda < 450$ | $2.76 < \Delta V < 4.0$ | Indium gallium nitride (InGaN) |
| Purple | Multiple types | $2.48 < \Delta V < 3.7$ | Dual blue/red LEDs, blue with red phosphor, or white with purple plastic |
| Ultraviolet | $\lambda < 400$ | $3 < \Delta V < 4.1$ | Indium gallium nitride (InGaN) (385-400 nm) Diamond (235 nm) Boron nitride (215 nm) Aluminium nitride (AlN) (210 nm) Aluminium gallium nitride (AlGaN) Aluminium gallium indium nitride (AlGaInN)—down to 210 nm |
| Pink | Multiple types | $\Delta V \sim 3.3$ | Blue with one or two phosphor layers, yellow with red, orange or pink phosphor added afterwards, white with pink plastic, or white phosphors with pink pigment or dye over top. |
| White | Broad spectrum | $2.8 < \Delta V < 4.2$ | **Cool / Pure White:** Blue/UV diode with yellow phosphor **Warm White:** Blue diode with orange phosphor |

## Blue and Ultraviolet

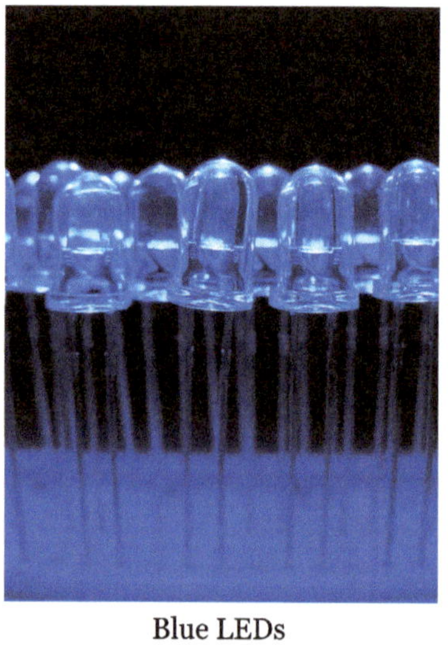

Blue LEDs

The first blue-violet LED using magnesium-doped gallium nitride was made at Stanford University in 1972 by Herb Maruska and Wally Rhines, doctoral students in materials science and engineering. At the time Maruska was on leave from RCA Laboratories, where he collaborated with Jacques Pankove on related work. In 1971, the year after Maruska left for Stanford, his RCA colleagues

Pankove and Ed Miller demonstrated the first blue electroluminescence from zinc-doped gallium nitride, though the subsequent device Pankove and Miller built, the first actual gallium nitride light-emitting diode, emitted green light. In 1974 the U.S. Patent Office awarded Maruska, Rhines and Stanford professor David Stevenson a patent for their work in 1972 (U.S. Patent US3819974 A) and today, magnesium-doping of gallium nitride remains the basis for all commercial blue LEDs and laser diodes. In the early 1970s, these devices were too dim for practical use, and research into gallium nitride devices slowed. In August 1989, Cree introduced the first commercially available blue LED based on the indirect bandgap semiconductor, silicon carbide (SiC). SiC LEDs had very low efficiency, no more than about 0.03%, but did emit in the blue portion of the visible light spectrum.

In the late 1980s, key breakthroughs in GaN epitaxial growth and p-type doping ushered in the modern era of GaN-based optoelectronic devices. Building upon this foundation, Theodore Moustakas at Boston University patented a method for producing high-brightness blue LEDs using a new two-step process. Two years later, in 1993, high-brightness blue LEDs were demonstrated again by Shuji Nakamura of Nichia Corporation using a gallium nitride growth process similar to Moustakas's. Both Moustakas and Nakamura were issued separate patents, which confused the issue of who was the original inventor (partly because although Moustakas invented his first, Nakamura filed first). This new development revolutionized LED lighting, making high-power blue light sources practical, leading to the development of technologies like Blu-ray, as well as allowing the bright high-resolution screens of modern tablets and phones.

Nakamura was awarded the 2006 Millennium Technology Prize for his invention. Nakamura, Hiroshi Amano and Isamu Akasaki were awarded the Nobel Prize in Physics in 2014 for the invention of the blue LED. In 2015, a US court ruled that three companies (i.e. the litigants who had not previously settled out of court) that had licensed Nakamura's patents for production in the United States had infringed Moustakas's prior patent, and ordered them to pay licensing fees of not less than 13 million USD.

By the late 1990s, blue LEDs became widely available. They have an active region consisting of one or more InGaN quantum wells sandwiched between thicker layers of GaN, called cladding layers. By varying the relative In/Ga fraction in the InGaN quantum wells, the light emission can in theory be varied from violet to amber. Aluminium gallium nitride (AlGaN) of varying Al/Ga fraction can be used to manufacture the cladding and quantum well layers for ultraviolet LEDs, but these devices have not yet reached the level of efficiency and technological maturity of InGaN/GaN blue/green devices. If un-alloyed GaN is used in this case to form the active quantum well layers, the device emits near-ultraviolet light with a peak wavelength centred around 365 nm. Green LEDs manufactured from the InGaN/GaN system are far more efficient and brighter than green LEDs produced with non-nitride material systems, but practical devices still exhibit efficiency too low for high-brightness applications.

With nitrides containing aluminium, most often AlGaN and AlGaInN, even shorter wavelengths are achievable. Ultraviolet LEDs in a range of wavelengths are becoming available on the market. Near-UV emitters at wavelengths around 375–395 nm are already cheap and often encountered, for example, as black light lamp replacements for inspection of anti-counterfeiting UV watermarks in some documents and paper currencies. Shorter-wavelength diodes, while substantially more expensive, are commercially available for wavelengths down to 240 nm. As the photosensitivity of microorganisms approximately matches the absorption spectrum of DNA, with a peak at about 260 nm, UV LED emitting at 250–270 nm are to be expected in prospective disinfection and sterilization devices. Recent research has shown that commercially available UVA LEDs (365 nm) are already effective disinfection and sterilization devices. UV-C wavelengths were obtained in laboratories using aluminium nitride (210 nm), boron nitride (215 nm) and diamond (235 nm).

## RGB

RGB-SMD-LED

RGB LEDs consist of one red, one green, and one blue LED. By independently adjusting each of the three, RGB LEDs are capable of producing a wide color gamut. Unlike dedicated-color LEDs, however, these obviously do not produce pure wavelengths. Moreover, such modules as commercially available are often not optimized for smooth color mixing.

## White

There are two primary ways of producing white light-emitting diodes (WLEDs), LEDs that generate high-intensity white light. One is to use individual LEDs that emit three primary colors—red, green, and blue—and then mix all the colors to form white light. The other is to use a phosphor material to convert monochromatic light from a blue or UV LED to broad-spectrum white light, much in the same way a fluorescent light bulb works. It is important to note that the 'whiteness' of the light produced is essentially engineered to suit the human eye, and depending on the situation it may not always be appropriate to think of it as white light.

There are three main methods of mixing colors to produce white light from an LED:

- blue LED + green LED + red LED (color mixing; can be used as backlighting for displays, extremely poor for illumination due to gaps in spectrum)
- near-UV or UV LED + RGB phosphor (an LED producing light with a wavelength shorter than blue's is used to excite an RGB phosphor)
- blue LED + yellow phosphor (two complementary colors combine to form white light; more efficient than first two methods and more commonly used)

Because of metamerism, it is possible to have quite different spectra that appear white. However, the appearance of objects illuminated by that light may vary as the spectrum varies, this is the issue of Colour rendition, quite separate from Colour Temperature, where a really orange or cyan object could appear with the wrong colour and much darker as the LED or phosphor does not emit the wavelength. The best colour rendition CFL and LEDs use a mix of phosphors, resulting in less efficiency but better quality of light. Though incandescent halogen lamps have a more orange colour temperature, they are still the best easily available artificial light sources in terms of colour rendition.

## RGB Systems

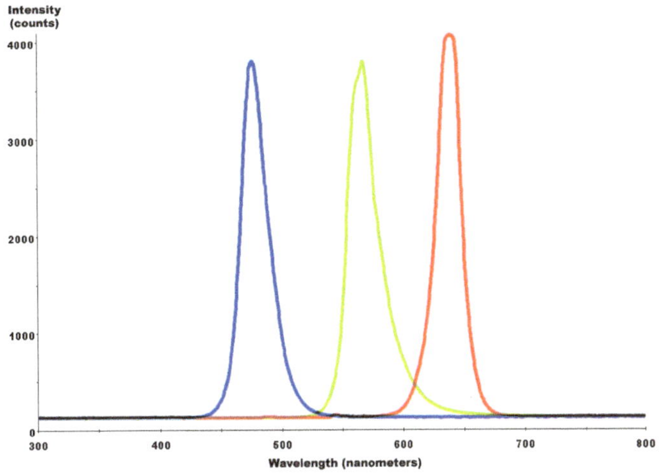


Combined spectral curves for blue, yellow-green, and high-brightness red solid-state semiconductor LEDs. FWHM spectral bandwidth is approximately 24–27 nm for all three colors

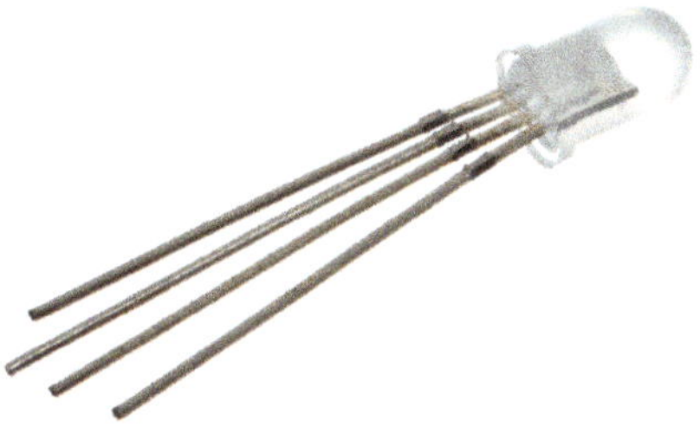

RGB LED

White light can be formed by mixing differently colored lights; the most common method is to use red, green, and blue (RGB). Hence the method is called multi-color white LEDs (sometimes referred to as RGB LEDs). Because these need electronic circuits to control the blending and diffusion of different colors, and because the individual color LEDs typically have slightly different emission patterns (leading to variation of the color depending on direction) even if they are made as a single unit, these are seldom used to produce white lighting. Nonetheless, this method has many applications because of the flexibility of mixing different colors, and in principle, this mechanism also has higher quantum efficiency in producing white light.

There are several types of multi-color white LEDs: di-, tri-, and tetrachromatic white LEDs. Several key factors that play among these different methods include color stability, color rendering capability, and luminous efficacy. Often, higher efficiency means lower color rendering, presenting a trade-off between the luminous efficacy and color rendering. For example, the dichromatic white LEDs have the best luminous efficacy (120 lm/W), but the lowest color rendering capability. However, although tetrachromatic white LEDs have excellent color rendering capability, they often have poor luminous efficacy. Trichromatic white LEDs are in between, having both good luminous efficacy (>70 lm/W) and fair color rendering capability.

One of the challenges is the development of more efficient green LEDs. The theoretical maximum for green LEDs is 683 lumens per watt but as of 2010 few green LEDs exceed even 100 lumens per watt. The blue and red LEDs get closer to their theoretical limits.

Multi-color LEDs offer not merely another means to form white light but a new means to form light of different colors. Most perceivable colors can be formed by mixing different amounts of three primary colors. This allows precise dynamic color control. As more effort is devoted to investigating this method, multi-color LEDs should have profound influence on the fundamental method that we use to produce and control light color. However, before this type of LED can play a role on the market, several technical problems must be solved. These include that this type of LED's emission power decays exponentially with rising temperature, resulting in a substantial change in color stability. Such problems inhibit and may preclude industrial use. Thus, many new package designs aimed at solving this problem have been proposed and their results are now being reproduced by researchers and scientists. However multi-colour LEDs without phosphors can never provide good quality lighting because each LED is a narrow band source. LEDs without phosphor while a poorer solution for general lighting are the best solution for displays, either backlight of LCD, or direct LED based pixels.

Correlated color temperature (CCT) dimming for LED technology is regarded as a difficult task since binning, age and temperature drift effects of LEDs change the actual color value output. Feedback loop systems are used for example with color sensors, to actively monitor and control the color output of multiple color mixing LEDs.

## Phosphor-based LEDs

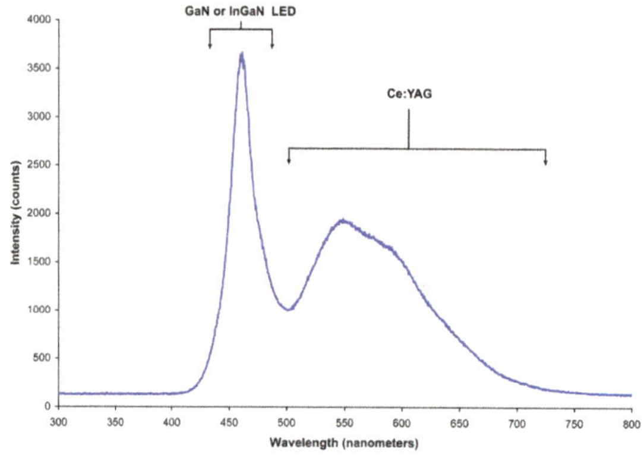


Spectrum of a white LED showing blue light directly emitted by the GaN-based LED (peak at about 465 nm) and the more broadband Stokes-shifted light emitted by the $Ce^{3+}$:YAG phosphor, which emits at roughly 500–700 nm

This method involves coating LEDs of one color (mostly blue LEDs made of InGaN) with phosphors of different colors to form white light; the resultant LEDs are called phosphor-based or phosphor-converted white LEDs (pcLEDs). A fraction of the blue light undergoes the Stokes shift being transformed from shorter wavelengths to longer. Depending on the color of the original LED, phosphors of different colors can be employed. If several phosphor layers of distinct colors are applied, the emitted spectrum is broadened, effectively raising the color rendering index (CRI) value of a given LED.

Phosphor-based LED efficiency losses are due to the heat loss from the Stokes shift and also other phosphor-related degradation issues. Their luminous efficacies compared to normal LEDs depend on the spectral distribution of the resultant light output and the original wavelength of the LED itself. For example, the luminous efficacy of a typical YAG yellow phosphor based white LED ranges from 3 to 5 times the luminous efficacy of the original blue LED because of the human eye's greater sensitivity to yellow than to blue (as modeled in the luminosity function). Due to the simplicity of manufacturing, the phosphor method is still the most popular method for making high-intensity white LEDs. The design and production of a light source or light fixture using a monochrome emitter with phosphor conversion is simpler and cheaper than a complex RGB system, and the majority of high-intensity white LEDs presently on the market are manufactured using phosphor light conversion.

Among the challenges being faced to improve the efficiency of LED-based white light sources is the development of more efficient phosphors. As of 2010, the most efficient yellow phosphor is still the YAG phosphor, with less than 10% Stokes shift loss. Losses attributable to internal optical losses due to re-absorption in the LED chip and in the LED packaging itself account typically for another 10% to 30% of efficiency loss. Currently, in the area of phosphor LED development, much effort is being spent on optimizing these devices to higher light output and higher operation temperatures. For instance, the efficiency can be raised by adapting better package design or by using a more suitable type of phosphor. Conformal coating process is frequently used to address the issue of varying phosphor thickness.

Some phosphor-based white LEDs encapsulate InGaN blue LEDs inside phosphor-coated epoxy. Alternatively, the LED might be paired with a remote phosphor, a preformed polycarbonate piece coated with the phosphor material. Remote phosphors provide more diffuse light, which is desirable for many applications. Remote phosphor designs are also more tolerant of variations in the LED emissions spectrum. A common yellow phosphor material is cerium-doped yttrium aluminium garnet ($Ce^{3+}$:YAG).

White LEDs can also be made by coating near-ultraviolet (NUV) LEDs with a mixture of high-efficiency europium-based phosphors that emit red and blue, plus copper and aluminium-doped zinc sulfide (ZnS:Cu, Al) that emits green. This is a method analogous to the way fluorescent lamps work. This method is less efficient than blue LEDs with YAG:Ce phosphor, as the Stokes shift is larger, so more energy is converted to heat, but yields light with better spectral characteristics, which render color better. Due to the higher radiative output of the ultraviolet LEDs than of the blue ones, both methods offer comparable brightness. A concern is that UV light may leak from a malfunctioning light source and cause harm to human eyes or skin.

## Other White LEDs

Another method used to produce experimental white light LEDs used no phosphors at all and was based on homoepitaxially grown zinc selenide (ZnSe) on a ZnSe substrate that simultaneously emitted blue light from its active region and yellow light from the substrate.

A new style of wafers composed of gallium-nitride-on-silicon (GaN-on-Si) is being used to produce white LEDs using 200-mm silicon wafers. This avoids the typical costly sapphire substrate in relatively small 100- or 150-mm wafer sizes. The sapphire apparatus must be coupled with a mirror-like collector to reflect light that would otherwise be wasted. It is predicted that by 2020, 40% of all GaN LEDs will be made with GaN-on-Si. Manufacturing large sapphire material is difficult, while large silicon material is cheaper and more abundant. LED companies shifting from using sapphire to silicon should be a minimal investment.

## Organic Light-emitting Diodes (OLEDs)

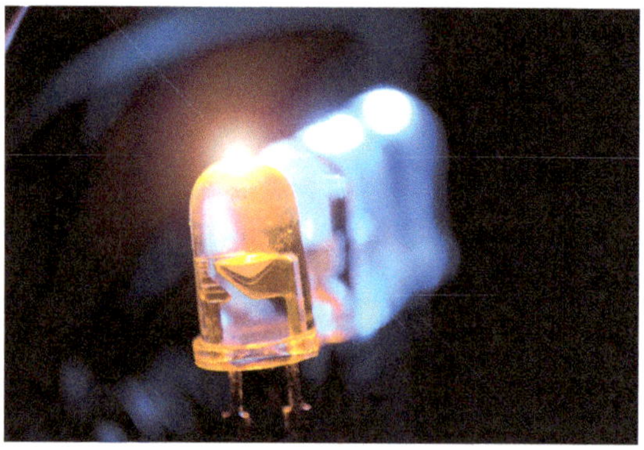

Orange light-emitting diode

In an organic light-emitting diode (OLED), the electroluminescent material comprising the emissive layer of the diode is an organic compound. The organic material is electrically conductive due to the delocalization of pi electrons caused by conjugation over all or part of the molecule, and the material therefore functions as an organic semiconductor. The organic materials can be small organic molecules in a crystalline phase, or polymers.

The potential advantages of OLEDs include thin, low-cost displays with a low driving voltage, wide viewing angle, and high contrast and color gamut. Polymer LEDs have the added benefit of printable and flexible displays. OLEDs have been used to make visual displays for portable electronic devices such as cellphones, digital cameras, and MP3 players while possible future uses include lighting and televisions.

## Quantum dot LEDs

Quantum dots (QD) are semiconductor nanocrystals with optical properties that let their emission color be tuned from the visible into the infrared spectrum. This allows quantum dot LEDs to create almost any color on the CIE diagram. This provides more color options and better color rendering than white LEDs since the emission spectrum is much narrower, characteristic of quantum confined states.

There are two types of schemes for QD excitation. One uses photo excitation with a primary light source LED (typically blue or UV LEDs are used). The other is direct electrical excitation first demonstrated by Alivisatos et al.

One example of the photo-excitation scheme is a method developed by Michael Bowers, at Vanderbilt University in Nashville, involving coating a blue LED with quantum dots that glow white in response to the blue light from the LED. This method emits a warm, yellowish-white light similar to that made by incandescent light bulbs. Quantum dots are also being considered for use in white light-emitting diodes in liquid crystal display (LCD) televisions.

In February 2011 scientists at PlasmaChem GmbH were able to synthesize quantum dots for LED applications and build a light converter on their basis, which was able to efficiently convert light from blue to any other color for many hundred hours. Such QDs can be used to emit visible or near infrared light of any wavelength being excited by light with a shorter wavelength.

The structure of QD-LEDs used for the electrical-excitation scheme is similar to basic design of OLEDs. A layer of quantum dots is sandwiched between layers of electron-transporting and hole-transporting materials. An applied electric field causes electrons and holes to move into the quantum dot layer and recombine forming an exciton that excites a QD. This scheme is commonly studied for quantum dot display. The tunability of emission wavelengths and narrow bandwidth is also beneficial as excitation sources for fluorescence imaging. Fluorescence near-field scanning optical microscopy (NSOM) utilizing an integrated QD-LED has been demonstrated.

In February 2008, a luminous efficacy of 300 lumens of visible light per watt of radiation (not per electrical watt) and warm-light emission was achieved by using nanocrystals.

## Types

LEDs are produced in a variety of shapes and sizes. The color of the plastic lens is often the same as the actual color of light emitted, but not always. For instance, purple plastic is often used for infrared LEDs, and most blue devices have colorless housings. Modern high-power LEDs such as those used for lighting and backlighting are generally found in surface-mount technology (SMT) packages (not shown).

The main types of LEDs are miniature, high-power devices and custom designs such as alphanumeric or multi-color.

### Miniature

These are mostly single-die LEDs used as indicators, and they come in various sizes from 2 mm to 8 mm, through-hole and surface mount packages. They usually do not use a separate heat sink. Typical current ratings range from around 1 mA to above 20 mA. The small size sets a natural upper boundary on power consumption due to heat caused by the high current density and need for a heat sink. Often daisy chained as used in LED tapes.

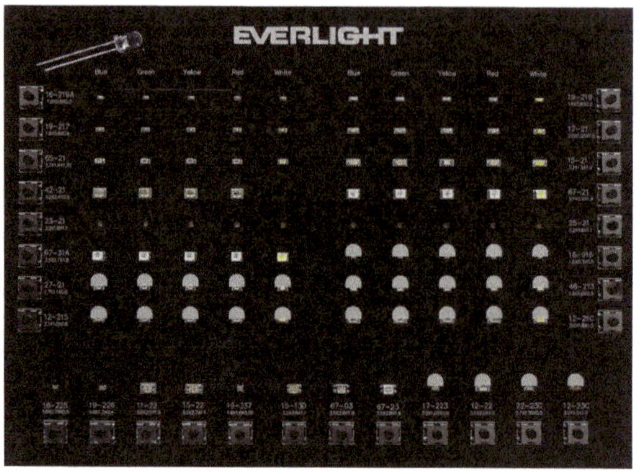


Photo of miniature surface mount LEDs in most common sizes. They can be much smaller than a traditional 5 mm lamp type LED, shown on the upper left corner

Common package shapes include round, with a domed or flat top, rectangular with a flat top (as used in bar-graph displays), and triangular or square with a flat top. The encapsulation may also be clear or tinted to improve contrast and viewing angle.

Very small (1.6x1.6x0.35 mm) red, green, and blue surface mount miniature LED package with gold wire bonding details

Researchers at the University of Washington have invented the thinnest LED. It is made of two-dimensional (2-D) flexible materials. It is three atoms thick, which is 10 to 20 times thinner than three-dimensional (3-D) LEDs and is also 10,000 times smaller than the thickness of a human hair. These 2-D LEDs are going to make it possible to create smaller, more energy-efficient lighting, optical communication and nano lasers.

There are three main categories of miniature single die LEDs:

Low-current

Typically rated for 2 mA at around 2 V (approximately 4 mW consumption)

Standard

20 mA LEDs (ranging from approximately 40 mW to 90 mW) at around:

- 1.9 to 2.1 V for red, orange, yellow, and traditional green
- 3.0 to 3.4 V for pure green and blue
- 2.9 to 4.2 V for violet, pink, purple and white

Ultra-high-output

20 mA at approximately 2 or 4–5 V, designed for viewing in direct sunlight

5 V and 12 V LEDs are ordinary miniature LEDs that incorporate a suitable series resistor for direct connection to a 5 V or 12 V supply.

## High-power

High-power light-emitting diodes attached to an LED star base (Luxeon, Lumileds)

High-power LEDs (HP-LEDs) or high-output LEDs (HO-LEDs) can be driven at currents from hundreds of mA to more than an ampere, compared with the tens of mA for other LEDs. Some can emit over a thousand lumens. LED power densities up to 300 W/cm$^2$ have been achieved. Since overheating is destructive, the HP-LEDs must be mounted on a heat sink to allow for heat dissipation. If the heat from an HP-LED is not removed, the device fails in seconds. One HP-LED can often replace an incandescent bulb in a flashlight, or be set in an array to form a powerful LED lamp.

Some well-known HP-LEDs in this category are the Nichia 19 series, Lumileds Rebel Led, Osram Opto Semiconductors Golden Dragon, and Cree X-lamp. As of September 2009, some HP-LEDs manufactured by Cree now exceed 105 lm/W.

Examples for Haitz's law—which predicts an exponential rise in light output and efficacy of LEDs over time—are the CREE XP-G series LED, which achieved 105 lm/W in 2009 and the Nichia 19 series with a typical efficacy of 140 lm/W, released in 2010.

## AC Driven

LEDs developed by Seoul Semiconductor can operate on AC power without a DC converter. For each half-cycle, part of the LED emits light and part is dark, and this is reversed during the next half-cycle. The efficacy of this type of HP-LED is typically 40 lm/W. A large number of LED elements in series may be able to operate directly from line voltage. In 2009, Seoul Semiconductor released a high DC voltage LED, named as 'Acrich MJT', capable of being driven from AC power with a simple controlling circuit. The low-power dissipation of these LEDs affords them more flexibility than the original AC LED design.

## Application-specific Variations

### Flashing

Flashing LEDs are used as attention seeking indicators without requiring external electronics. Flashing LEDs resemble standard LEDs but they contain an integrated multivibrator circuit that causes the LED to flash with a typical period of one second. In diffused lens LEDs, this circuit is visible as a small black dot. Most flashing LEDs emit light of one color, but more sophisticated devices can flash between multiple colors and even fade through a color sequence using RGB color mixing.

### Bi-color

Bi-color LEDs contain two different LED emitters in one case. There are two types of these. One type consists of two dies connected to the same two leads antiparallel to each other. Current flow in one direction emits one color, and current in the opposite direction emits the other color. The other type consists of two dies with separate leads for both dies and another lead for common anode or cathode so that they can be controlled independently. The most common bi-color combination is red/traditional green, however, other available combinations include amber/traditional green, red/pure green, red/blue, and blue/pure green.

### Tri-color

Tri-color LEDs contain three different LED emitters in one case. Each emitter is connected to a separate lead so they can be controlled independently. A four-lead arrangement is typical with one common lead (anode or cathode) and an additional lead for each color.

### RGB

RGB LEDs are tri-color LEDs with red, green, and blue emitters, in general using a four-wire connection with one common lead (anode or cathode). These LEDs can have either common positive or common negative leads. Others, however, have only two leads (positive and negative) and have a built-in tiny electronic control unit.

### Decorative-multicolor

Decorative-multicolor LEDs incorporate several emitters of different colors supplied by only two lead-out wires. Colors are switched internally by varying the supply voltage.

### Alphanumeric

Alphanumeric LEDs are available in seven-segment, starburst, and dot-matrix format. Seven-segment displays handle all numbers and a limited set of letters. Starburst displays can display all letters. Dot-matrix displays typically use 5x7 pixels per character. Seven-segment LED displays were in widespread use in the 1970s and 1980s, but rising use of liquid crystal displays, with their lower power needs and greater display flexibility, has reduced the popularity of numeric and alphanumeric LED displays.

### Digital-RGB

Digital-RGB LEDs are RGB LEDs that contain their own "smart" control electronics. In addition to power and ground, these provide connections for data-in, data-out, and sometimes a clock or strobe signal. These are connected in a daisy chain, with the data in of the first LED sourced by a microprocessor, which can control the brightness and color of each LED independently of the others. They are used where a combination of maximum control and minimum visible electronics are needed such as strings for Christmas and LED matrices. Some even have refresh rates in the kHz range, allowing for basic video applications.

### Filament

An LED filament consists of multiple LED chips connected in series on a common longitudinal substrate that forms a thin rod reminiscent of a traditional incandescent filament. These are being used as a low-cost decorative alternative for traditional light bulbs that are being phased out in many countries. The filaments require a rather high voltage to light to nominal brightness, allowing them to work efficiently and simply with mains voltages. Often a simple rectifier and capacitive current limiting are employed to create a low-cost replacement for a traditional light bulb without the complexity of the low voltage, high current converter that single die LEDs need. Usually, they are packaged in a sealed enclosure with a shape similar to lamps they were designed to replace (e.g. a bulb) and filled with inert nitrogen or carbon dioxide gas to remove heat efficiently.

## Considerations for use

### Power Sources

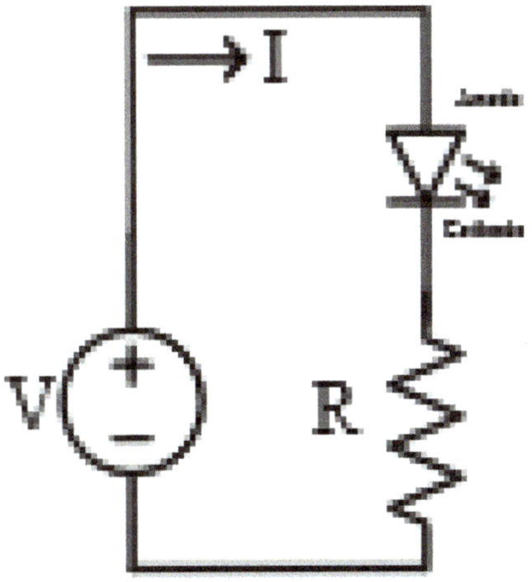


Simple LED circuit with resistor for current limiting

The current–voltage characteristic of an LED is similar to other diodes, in that the current is dependent exponentially on the voltage. This means that a small change in voltage can cause a large change in current. If the applied voltage exceeds the LED's forward voltage drop by a small amount, the current rating may be exceeded by a large amount, potentially damaging or destroying the LED. The typical solution is to use constant-current power supplies to keep the current below the LED's maximum current rating. Since most common power sources (batteries, mains) are constant-voltage sources, most LED fixtures must include a power converter, at least a current-limiting resistor. However, the high resistance of three-volt coin cells combined with the high differential resistance of nitride-based LEDs makes it possible to power such an LED from such a coin cell without an external resistor.

## Electrical Polarity

As with all diodes, current flows easily from p-type to n-type material. However, no current flows and no light is emitted if a small voltage is applied in the reverse direction. If the reverse voltage grows large enough to exceed the breakdown voltage, a large current flows and the LED may be damaged. If the reverse current is sufficiently limited to avoid damage, the reverse-conducting LED is a useful noise diode.

## Safety and Health

The vast majority of devices containing LEDs are "safe under all conditions of normal use", and so are classified as "Class 1 LED product"/"LED Klasse 1". At present, only a few LEDs—extremely bright LEDs that also have a tightly focused viewing angle of 8° or less—could, in theory, cause temporary blindness, and so are classified as "Class 2". The opinion of the French Agency for Food, Environmental and Occupational Health & Safety (ANSES) of 2010, on the health issues concerning LEDs, suggested banning public use of lamps in the moderate Risk Group 2, especially those with a high blue component, in places frequented by children.

In general, laser safety regulations—and the "Class 1", "Class 2", etc. system—also apply to LEDs.

While LEDs have the advantage over fluorescent lamps that they do not contain mercury, they may contain other hazardous metals such as lead and arsenic. Regarding the toxicity of LEDs when treated as waste, a study published in 2011 stated: "According to federal standards, LEDs are not hazardous except for low-intensity red LEDs, which leached Pb [lead] at levels exceeding regulatory limits (186 mg/L; regulatory limit: 5). However, according to California regulations, excessive levels of copper (up to 3892 mg/kg; limit: 2500), lead (up to 8103 mg/kg; limit: 1000), nickel (up to 4797 mg/kg; limit: 2000), or silver (up to 721 mg/kg; limit: 500) render all except low-intensity yellow LEDs hazardous."

In 2016 a statement of the American Medical Association (AMA) concerning the possible influence of blueish street lighting on the sleep-wake cycle of city-dwellers led to some controversy. So far high-pressure sodium lamps (HPS) with an orange light spectrum were the most efficient light sources commonly used in street-lighting. Now many modern street lamps are equipped with Indium gallium nitride LEDs (InGaN). These are even more efficient and mostly emit blue-rich light with a higher *correlated color temperature (CCT)*. Since light with a high CCT resembles daylight it is thought that this might have an effect on the normal circadian physiology by suppressing melatonin production in the human body. There have been no relevant studies as yet and critics claim exposure levels are not high enough to have a noticeable effect.

## Advantages

- Efficiency: LEDs emit more lumens per watt than incandescent light bulbs. The efficiency of LED lighting fixtures is not affected by shape and size, unlike fluorescent light bulbs or tubes.
- Color: LEDs can emit light of an intended color without using any color filters as traditional lighting methods need. This is more efficient and can lower initial costs.
- Size: LEDs can be very small (smaller than 2 $mm^2$) and are easily attached to printed circuit boards.
- Warmup time: LEDs light up very quickly. A typical red indicator LED achieves full brightness in under a microsecond. LEDs used in communications devices can have even faster response times.
- Cycling: LEDs are ideal for uses subject to frequent on-off cycling, unlike incandescent and fluorescent lamps that fail faster when cycled often, or high-intensity discharge lamps (HID lamps) that require a long time before restarting.
- Dimming: LEDs can very easily be dimmed either by pulse-width modulation or lowering the forward current. This pulse-width modulation is why LED lights, particularly headlights on cars, when viewed on camera or by some people, appear to be flashing or flickering. This is a type of stroboscopic effect.
- Cool light: In contrast to most light sources, LEDs radiate very little heat in the form of IR that can cause damage to sensitive objects or fabrics. Wasted energy is dispersed as heat through the base of the LED.
- Slow failure: LEDs mostly fail by dimming over time, rather than the abrupt failure of incandescent bulbs.
- Lifetime: LEDs can have a relatively long useful life. One report estimates 35,000 to 50,000 hours of useful life, though time to complete failure may be longer. Fluorescent tubes typically are rated at about 10,000 to 15,000 hours, depending partly on the conditions of use, and incandescent light bulbs at 1,000 to 2,000 hours. Several DOE demonstrations have shown that reduced maintenance costs from this extended lifetime, rather than energy savings, is the primary factor in determining the payback period for an LED product.
- Shock resistance: LEDs, being solid-state components, are difficult to damage with external shock, unlike fluorescent and incandescent bulbs, which are fragile.
- Focus: The solid package of the LED can be designed to focus its light. Incandescent and fluorescent sources often require an external reflector to collect light and direct it in a usable manner. For larger LED packages total internal reflection (TIR) lenses are often used to the same effect. However, when large quantities of light are needed many light sources are usually deployed, which are difficult to focus or collimate towards the same target.

## Disadvantages

- Initial price: LEDs are currently slightly more expensive (price per lumen) on an initial capital cost basis, than other lighting technologies. As of March 2014, at least one manufacturer claims to have reached $1 per kilolumen. The additional expense partially stems from the relatively low lumen output and the drive circuitry and power supplies needed.

- Temperature dependence: LED performance largely depends on the ambient temperature of the operating environment – or thermal management properties. Overdriving an LED in high ambient temperatures may result in overheating the LED package, eventually leading to device failure. An adequate heat sink is needed to maintain long life. This is especially important in automotive, medical, and military uses where devices must operate over a wide range of temperatures, which require low failure rates. Toshiba has produced LEDs with an operating temperature range of −40 to 100 °C, which suits the LEDs for both indoor and outdoor use in applications such as lamps, ceiling lighting, street lights, and floodlights.
- Voltage sensitivity: LEDs must be supplied with a voltage above their threshold voltage and a current below their rating. Current and lifetime change greatly with a small change in applied voltage. They thus require a current-regulated supply (usually just a series resistor for indicator LEDs).
- Color rendition: Most cool-white LEDs have spectra that differ significantly from a black body radiator like the sun or an incandescent light. The spike at 460 nm and dip at 500 nm can cause the color of objects to be perceived differently under cool-white LED illumination than sunlight or incandescent sources, due to metamerism, red surfaces being rendered particularly poorly by typical phosphor-based cool-white LEDs.
- Area light source: Single LEDs do not approximate a point source of light giving a spherical light distribution, but rather a lambertian distribution. So LEDs are difficult to apply to uses needing a spherical light field; however, different fields of light can be manipulated by the application of different optics or "lenses". LEDs cannot provide divergence below a few degrees. In contrast, lasers can emit beams with divergences of 0.2 degrees or less.
- Electrical polarity: Unlike incandescent light bulbs, which illuminate regardless of the electrical polarity, LEDs only light with correct electrical polarity. To automatically match source polarity to LED devices, rectifiers can be used.
- Blue hazard: There is a concern that blue LEDs and cool-white LEDs are now capable of exceeding safe limits of the so-called blue-light hazard as defined in eye safety specifications such as ANSI/IESNA RP-27.1–05: Recommended Practice for Photobiological Safety for Lamp and Lamp Systems.
- Light pollution: Because white LEDs, especially those with high color temperature, emit much more short wavelength light than conventional outdoor light sources such as high-pressure sodium vapor lamps, the increased blue and green sensitivity of scotopic vision means that white LEDs used in outdoor lighting cause substantially more sky glow. The American Medical Association warned on the use of high blue content white LEDs in street lighting, due to their higher impact on human health and environment, compared to low blue content light sources (e.g. High-Pressure Sodium, PC amber LEDs, and low CCT LEDs).

- Efficiency droop: The efficiency of LEDs decreases as the electric current increases. Heating also increases with higher currents, which compromises LED lifetime. These effects put practical limits on the current through an LED in high power applications.
- Impact on insects: LEDs are much more attractive to insects than sodium-vapor lights, so much so that there has been speculative concern about the possibility of disruption to food webs.
- Use in winter conditions: Since they do not give off much heat in comparison to incandescent lights, LED lights used for traffic control can have snow obscuring them, leading to accidents.

## Applications

LED uses fall into four major categories:

- Visual signals where light goes more or less directly from the source to the human eye, to convey a message or meaning
- Illumination where light is reflected from objects to give visual response of these objects
- Measuring and interacting with processes involving no human vision
- Narrow band light sensors where LEDs operate in a reverse-bias mode and respond to incident light, instead of emitting light

## Indicators and Signs

The low energy consumption, low maintenance and small size of LEDs has led to uses as status indicators and displays on a variety of equipment and installations. Large-area LED displays are used as stadium displays, dynamic decorative displays, and dynamic message signs on freeways. Thin, lightweight message displays are used at airports and railway stations, and as destination displays for trains, buses, trams, and ferries.

Red and green LED traffic signals

One-color light is well suited for traffic lights and signals, exit signs, emergency vehicle lighting, ships' navigation lights or lanterns (chromacity and luminance standards being set under the Convention on the International Regulations for Preventing Collisions at Sea 1972, Annex I and the CIE) and LED-based Christmas lights. In cold climates, LED traffic lights may remain snow-covered. Red or yellow LEDs are used in indicator and alphanumeric displays in environments where night vision must be retained: aircraft cockpits, submarine and ship bridges, astronomy observatories, and in the field, e.g. night time animal watching and military field use.

Automotive applications for LEDs continue to grow

Because of their long life, fast switching times, and visibility in broad daylight due to their high output and focus, LEDs have been used in brake lights for cars' high-mounted brake lights, trucks, and buses, and in turn signals for some time. However, many vehicles now use LEDs for their rear light clusters. The use in brakes improves safety, due to a great reduction in the time needed to light fully, or faster rise time, up to 0.5 second faster than an incandescent bulb. This gives drivers behind more time to react. In a dual intensity circuit (rear markers and brakes) if the LEDs are not pulsed at a fast enough frequency, they can create a phantom array, where ghost images of the LED appear if the eyes quickly scan across the array. White LED headlamps are beginning to appear. Using LEDs has styling advantages because LEDs can form much thinner lights than incandescent lamps with parabolic reflectors.

Due to the relative cheapness of low output LEDs, they are also used in many temporary uses such as glowsticks, throwies, and the photonic textile Lumalive. Artists have also used LEDs for LED art.

Weather and all-hazards radio receivers with Specific Area Message Encoding (SAME) have three LEDs: red for warnings, orange for watches, and yellow for advisories and statements whenever issued.

## Lighting

With the development of high-efficiency and high-power LEDs, it has become possible to use LEDs in lighting and illumination. To encourage the shift to LED lamps and other high-efficiency lighting, the US Department of Energy has created the L Prize competition. The Philips Lighting North America LED bulb won the first competition on August 3, 2011, after successfully completing 18 months of intensive field, lab, and product testing.

LEDs are used as street lights and in other architectural lighting. The mechanical robustness and long lifetime are used in automotive lighting on cars, motorcycles, and bicycle lights. LED light emission may be efficiently controlled by using nonimaging optics principles.

LED street lights are employed on poles and in parking garages. In 2007, the Italian village of Torraca was the first place to convert its entire illumination system to LEDs.

LEDs are used in aviation lighting. Airbus has used LED lighting in its Airbus A320 Enhanced since 2007, and Boeing uses LED lighting in the 787. LEDs are also being used now in airport and heliport lighting. LED airport fixtures currently include medium-intensity runway lights, runway centerline lights, taxiway centerline and edge lights, guidance signs, and obstruction lighting.

LEDs are also used as a light source for DLP projectors, and to backlight LCD televisions (referred to as LED TVs) and laptop displays. RGB LEDs raise the color gamut by as much as 45%. Screens for TV and computer displays can be made thinner using LEDs for backlighting.

The lack of IR or heat radiation makes LEDs ideal for stage lights using banks of RGB LEDs that can easily change color and decrease heating from traditional stage lighting, as well as medical lighting where IR-radiation can be harmful. In energy conservation, the lower heat output of LEDs also means air conditioning (cooling) systems have less heat in need of disposal.

LEDs are small, durable and need little power, so they are used in handheld devices such as flashlights. LED strobe lights or camera flashes operate at a safe, low voltage, instead of the 250+ volts commonly found in xenon flashlamp-based lighting. This is especially useful in cameras on mobile phones, where space is at a premium and bulky voltage-raising circuitry is undesirable.

LEDs are used for infrared illumination in night vision uses including security cameras. A ring of LEDs around a video camera, aimed forward into a retroreflective background, allows chroma keying in video productions.

LED for miners, to increase visibility inside mines

LEDs are used in mining operations, as cap lamps to provide light for miners. Research has been done to improve LEDs for mining, to reduce glare and to increase illumination, reducing risk of injury to the miners.

LEDs are now used commonly in all market areas from commercial to home use: standard lighting, AV, stage, theatrical, architectural, and public installations, and wherever artificial light is used.

LEDs are increasingly finding uses in medical and educational applications, for example as mood enhancement, and new technologies such as AmBX, exploiting LED versatility. NASA has even sponsored research for the use of LEDs to promote health for astronauts.

## Data Communication and Other Signalling

Light can be used to transmit data and analog signals. For example, lighting white LEDs can be used in systems assisting people to navigate in closed spaces while searching necessary rooms or objects.

Assistive listening devices in many theaters and similar spaces use arrays of infrared LEDs to send sound to listeners' receivers. Light-emitting diodes (as well as semiconductor lasers) are used to send data over many types of fiber optic cable, from digital audio over TOSLINK cables to the very high bandwidth fiber links that form the Internet backbone. For some time, computers were commonly equipped with IrDA interfaces, which allowed them to send and receive data to nearby machines via infrared.

Because LEDs can cycle on and off millions of times per second, very high data bandwidth can be achieved.

## Sustainable Lighting

Efficient lighting is needed for sustainable architecture. In 2009, US Department of Energy testing results on LED lamps showed an average efficacy of 35 lm/W, below that of typical CFLs, and as low as 9 lm/W, worse than standard incandescent bulbs. A typical 13-watt LED lamp emitted 450 to 650 lumens, which is equivalent to a standard 40-watt incandescent bulb.

However, as of 2011, there are LED bulbs available as efficient as 150 lm/W and even inexpensive low-end models typically exceed 50 lm/W, so that a 6-watt LED could achieve the same results as a standard 40-watt incandescent bulb. The latter has an expected lifespan of 1,000 hours, whereas an LED can continue to operate with reduced efficiency for more than 50,000 hours.

See the chart below for a comparison of common light types:

| | **LED** | **CFL** | **Incandescent** |
|---|---|---|---|
| Lightbulb Projected Lifespan | 50,000 hours | 10,000 hours | 1,200 hours |
| Watts Per Bulb (equiv. 60 watts) | 10 | 14 | 60 |
| Cost Per Bulb | $2.00 | $7.00 | $1.25 |
| kWh of Electricity Used Over 50,000 Hours | 500 | 700 | 3000 |
| Cost of Electricity (@ 0.10 per kWh) | $50 | $70 | $300 |
| Bulbs Needed for 50,000 Hours of Use | 1 | 5 | 42 |
| Equivalent 50,000 Hours Bulb Expense | $2.00 | $35.00 | $52.50 |
| TOTAL Cost for 50,000 Hours | $52.00 | $105.00 | $352.50 |

## Energy Consumption

In the US, one kilowatt-hour (3.6 MJ) of electricity currently causes an average 1.34 pounds (610 g) of $CO_2$ emission. Assuming the average light bulb is on for 10 hours a day, a 40-watt bulb causes 196 pounds (89 kg) of $CO_2$ emission per year. The 6-watt LED equivalent only causes 30 pounds (14 kg) of $CO_2$ over the same time span. A building's carbon footprint from lighting can, therefore, be reduced by 85% by exchanging all incandescent bulbs for new LEDs—if a building previously used only incandescent bulbs.

In practice, most buildings that use a lot of lighting use fluorescent lighting, which has 22% luminous efficiency compared with 5% for filaments, so changing to LED lighting would still give a 34% reduction in electrical power use and carbon emissions.

The reduction in carbon emissions depends on the source of electricity. Nuclear power in the United States produced 19.2% of electricity in 2011, so reducing electricity consumption in the U.S. reduces carbon emissions more than in France (75% nuclear electricity) or Norway (almost entirely hydroelectric).

Replacing lights that spend the most time lit results in the most savings, so LED lights in infrequently used locations bring a smaller return on investment.

## Light Sources for Machine Vision systems

Machine vision systems often require bright and homogeneous illumination, so features of interest are easier to process. LEDs are often used for this purpose, and this is likely to remain one of their major uses until the price drops low enough to make signaling and illumination uses more widespread. Barcode scanners are the most common example of machine vision, and many low-cost products use red LEDs instead of lasers. Optical computer mice are an example of LEDs in machine vision, as it is used to provide an even light source on the surface for the miniature camera within the mouse. LEDs constitute a nearly ideal light source for machine vision systems for several reasons:

- The size of the illuminated field is usually comparatively small and machine vision systems are often quite expensive, so the cost of the light source is usually a minor concern. However, it might not be easy to replace a broken light source placed within complex machinery, and here the long service life of LEDs is a benefit.
- LED elements tend to be small and can be placed with high density over flat or even-shaped substrates (PCBs etc.) so that bright and homogeneous sources that direct light from tightly controlled directions on inspected parts can be designed. This can often be obtained with small, low-cost lenses and diffusers, helping to achieve high light densities with control over lighting levels and homogeneity. LED sources can be shaped in several configurations (spot lights for reflective illumination; ring lights for coaxial illumination; backlights for contour illumination; linear assemblies; flat, large format panels; dome sources for diffused, omnidirectional illumination).
- LEDs can be easily strobed (in the microsecond range and below) and synchronized with imaging. High-power LEDs are available allowing well-lit images even with very short light pulses. This is often used to obtain crisp and sharp "still" images of quickly moving parts.
- LEDs come in several different colors and wavelengths, allowing easy use of the best color for each need, where different color may provide better visibility of features of interest. Having a precisely known spectrum lets tightly matched filters be used to separate informative bandwidth or to reduce disturbing effects of ambient light. LEDs usually operate at comparatively low working temperatures, simplifying heat management, and dissipation. This allows using plastic lenses, filters, and diffusers. Waterproof units can also easily be designed, allowing use in harsh or wet environments (food, beverage, oil industries).

| | | |
|---|---|---|
|  LED digital display that can display four digits and points |  Traffic light using LED |  LED daytime running lights of Audi A4 |

## Other Applications

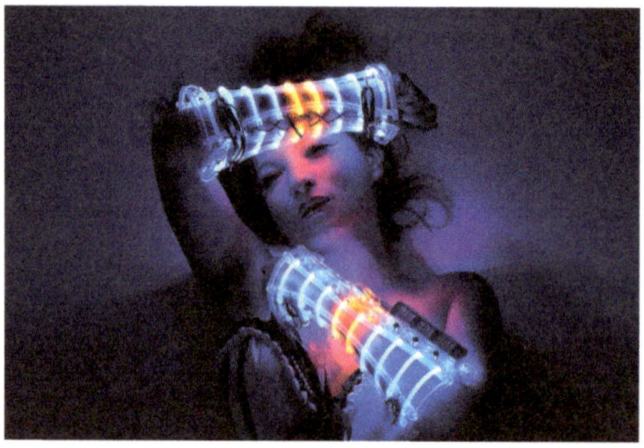

LED costume for stage performers

The light from LEDs can be modulated very quickly so they are used extensively in optical fiber and free space optics communications. This includes remote controls, such as for TVs, VCRs, and LED Computers, where infrared LEDs are often used. Opto-isolators use an LED combined with a photodiode or phototransistor to provide a signal path with electrical isolation between two circuits. This is especially useful in medical equipment where the signals from a low-voltage sensor circuit (usually battery-powered) in contact with a living organism must be electrically isolated from any possible electrical failure in a recording or monitoring device operating at potentially dangerous voltages. An optoisolator also lets information be transferred between circuits that don't share a common ground potential.

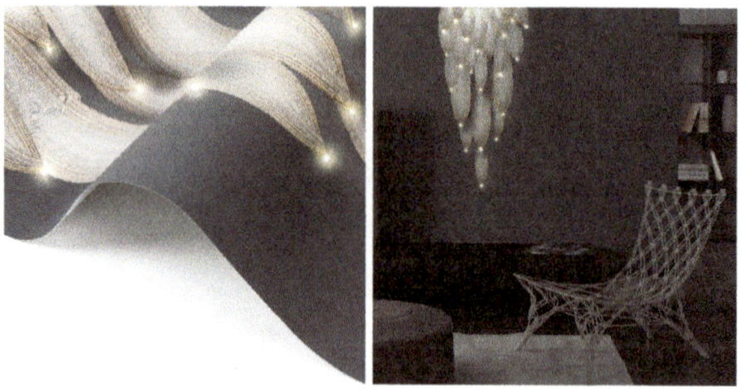

LED wallpaper by Meystyle

Many sensor systems rely on light as the signal source. LEDs are often ideal as a light source due to the requirements of the sensors. LEDs are used as motion sensors, for example in optical computer mice. The Nintendo Wii's sensor bar uses infrared LEDs. Pulse oximeters use them for measuring oxygen saturation. Some flatbed scanners use arrays of RGB LEDs rather than the typical cold-cathode fluorescent lamp as the light source. Having independent control of three illuminated colors allows the scanner to calibrate itself for more accurate color balance, and there is no need for warm-up. Further, its sensors only need be monochromatic, since at any one time the page being scanned is only lit by one color of light. Since LEDs can also be used as photodiodes, they can be used for both photo emission and detection. This could be used, for example, in a touchscreen that registers reflected light from a finger or stylus. Many materials and biological systems are sensitive to, or dependent on, light. Grow lights use LEDs to increase photosynthesis in plants, and bacteria and viruses can be removed from water and other substances using UV LEDs for sterilization.

LEDs have also been used as a medium-quality voltage reference in electronic circuits. The forward voltage drop (e.g. about 1.7 V for a normal red LED) can be used instead of a Zener diode in low-voltage regulators. Red LEDs have the flattest I/V curve above the knee. Nitride-based LEDs have a fairly steep I/V curve and are useless for this purpose. Although LED forward voltage is far more current-dependent than a Zener diode, Zener diodes with breakdown voltages below 3 V are not widely available.

The progressive miniaturization of low-voltage lighting technology, such as LEDs and OLEDs, suitable to incorporate into low-thickness materials has fostered experimentation in combining light sources and wall covering surfaces for interior walls. The new possibilities offered by these developments have prompted some designers and companies, such as Meystyle, Ingo Maurer, Lomox and Philips, to research and develop proprietary LED wallpaper technologies, some of which are currently available for commercial purchase. Other solutions mainly exist as prototypes or are in the process of being further refined.

## Integrated Circuit

Erasable Programmable Read-Only Memory (EPROM) integrated circuits. These packages have a transparent window that shows the die inside. The window allows the memory to be erased by exposing the chip to ultraviolet light

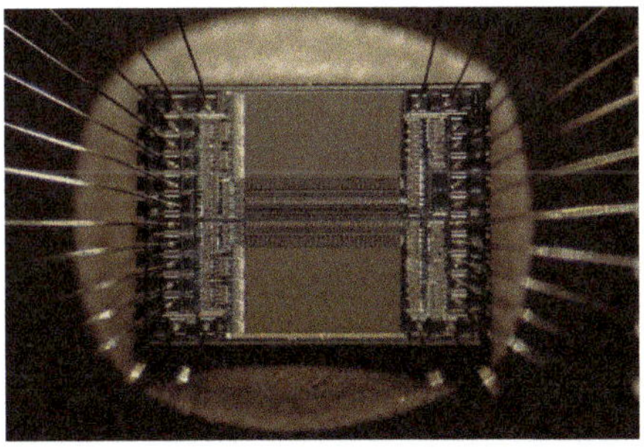

Integrated circuit from an EPROM memory microchip showing the memory blocks, the supporting circuitry and the fine silver wires which connect the integrated circuit die to the legs of the packaging

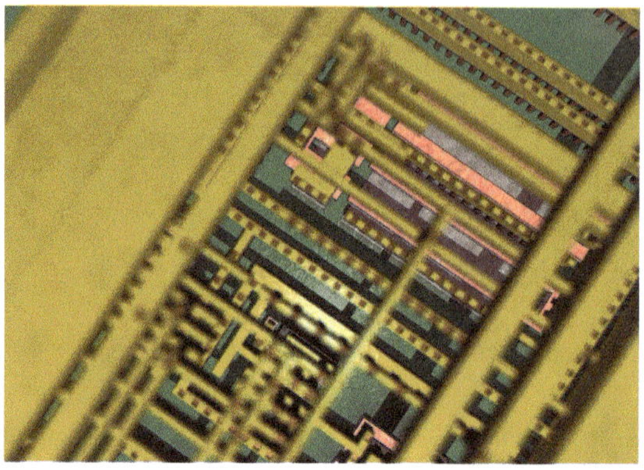

Synthetic detail of an integrated circuit through four layers of planarized copper interconnect, down to the polysilicon (pink), wells (greyish), and substrate (green)

An integrated circuit or monolithic integrated circuit (also referred to as an IC, a chip, or a microchip) is a set of electronic circuits on one small flat piece (or "chip") of semiconductor material, normally silicon. The integration of large numbers of tiny transistors into a small chip resulted in circuits that are orders of magnitude smaller, cheaper, and faster than those constructed of discrete electronic components. The IC's mass production capability, reliability and building-block approach to circuit design ensured the rapid adoption of standardized ICs in place of designs using discrete transistors. ICs are now used in virtually all electronic equipment and have revolutionized the world of electronics. Computers, mobile phones, and other digital home appliances are now inextricable parts of the structure of modern societies, made possible by the small size and low cost of ICs.

ICs were made possible by experimental discoveries showing that semiconductor devices could perform the functions of vacuum tubes, and by mid-20th-century technology advancements in semiconductor device fabrication. Since their origins in the 1960s, the size, speed, and capacity of chips have progressed enormously, driven by technical advances that allow more and more transistors on chips of the same size - a modern chip may have several billion transistors in an area the size of a human fingernail. These advances, roughly following Moore's law, allow a computer chip of 2016 to have millions of times the capacity and thousands of times the speed of the computer chips of the early 1970s.

ICs have two main advantages over discrete circuits: cost and performance. Cost is low because the chips, with all their components, are printed as a unit by photolithography rather than being constructed one transistor at a time. Furthermore, packaged ICs use much less material than discrete circuits. Performance is high because the IC's components switch quickly and consume little power (compared to their discrete counterparts) because of their small size and close proximity. The main disadvantage of ICs is the high cost to design them and fabricate the required photomasks. This high initial cost means ICs are only practical when high production volumes are anticipated.

## Terminology

An *integrated circuit* is defined as:

A circuit in which all or some of the circuit elements are inseparably associated and electrically interconnected so that it is considered to be indivisible for the purposes of construction and commerce.

Circuits meeting this definition can be constructed using many different technologies, including thin-film transistor, thick film technology, or hybrid integrated circuit. However, in general usage *integrated circuit* has come to refer to the single-piece circuit construction originally known as a *monolithic integrated circuit.*

## Invention

Early developments of the integrated circuit go back to 1949, when German engineer Werner Jacobi (Siemens AG) filed a patent for an integrated-circuit-like semiconductor amplifying device showing five transistors on a common substrate in a 3-stage amplifier arrangement. Jacobi disclosed small and cheap hearing aids as typical industrial applications of his patent. An immediate commercial use of his patent has not been reported.

The idea of the integrated circuit was conceived by Geoffrey Dummer (1909–2002), a radar scientist working for the Royal Radar Establishment of the British Ministry of Defence. Dummer presented the idea to the public at the Symposium on Progress in Quality Electronic Components in Washington, D.C. on 7 May 1952. He gave many symposia publicly to propagate his ideas, and unsuccessfully attempted to build such a circuit in 1956.

A precursor idea to the IC was to create small ceramic squares (wafers), each containing a single miniaturized component. Components could then be integrated and wired into a bidimensional or tridimensional compact grid. This idea, which seemed very promising in 1957, was proposed to the US Army by Jack Kilby and led to the short-lived Micromodule Program (similar to 1951's Project Tinkertoy). However, as the project was gaining momentum, Kilby came up with a new, revolutionary design: the IC.

Jack Kilby's original integrated circuit

Newly employed by Texas Instruments, Kilby recorded his initial ideas concerning the integrated circuit in July 1958, successfully demonstrating the first working integrated example on 12 September 1958. In his patent application of 6 February 1959, Kilby described his new device as "a body of semiconductor material ... wherein all the components of the electronic circuit are completely integrated." The first customer for the new invention was the US Air Force.

Kilby won the 2000 Nobel Prize in Physics for his part in the invention of the integrated circuit. His work was named an IEEE Milestone in 2009.

Half a year after Kilby, Robert Noyce at Fairchild Semiconductor developed his own idea of an integrated circuit that solved many practical problems Kilby's had not. Noyce's design was made of silicon, whereas Kilby's chip was made of germanium. Noyce credited Kurt Lehovec of Sprague Electric for the principle of p–n junction isolation, a key concept behind the IC. This isolation allows each transistor to operate independently despite being parts of the same piece of silicon.

Fairchild Semiconductor was also home of the first silicon-gate IC technology with self-aligned gates, the basis of all modern CMOS computer chips. The technology was developed by Italian physicist Federico Faggin in 1968. In 1970, he joined Intel in order to develop the first single-chip central processing unit (CPU) microprocessor, the Intel 4004, for which he received the National Medal of Technology and Innovation in 2010. The 4004 was designed by Busicom's Masatoshi Shima and Intel's Ted Hoff in 1969, but it was Faggin's improved design in 1970 that made it a reality.

## Advances

Advances in IC technology, primarily smaller features and larger chips, have allowed the number of transistors in an integrated circuit to double every two years, a trend known as Moore's law. This increased capacity has been used to decrease cost and increase functionality. In general, as the feature size shrinks, almost every aspect of an IC's operation improves. The cost per transistor and the switching power consumption per transistor go down, while the memory capacity and speed go up, through the relationships defined by Dennard scaling. Since these speed, capacity, and power consumption gains are apparent to the end user, there is fierce competition among the manufacturers to use finer geometries. Over the years, transistor sizes have decreased from 10s of microns in the early 1970s to 10 nanometers in 2017 with a corresponding million-fold increase in transistors per unit area. As of 2016, typical chip areas range from a few square millimeters to around 600 $mm^2$, with up to 25 million transistors per $mm^2$.

The expected shrinking of feature sizes, and the needed progress in related areas, was forecast for many years by the International Technology Roadmap for Semiconductors (ITRS). The final ITRS was issued in 2016, and it is being replaced by the International Roadmap for Devices and Systems.

Initially, ICs were strictly electronic devices. The success of ICs has led to the integration of other technologies, in the attempt to obtain the same advantages of small size and low cost. These technologies include mechanical devices, optics, and sensors.

- Very small mechanical devices driven by electricity can be integrated onto chips, a technology known as microelectromechanical systems. These devices were developed in the late 1980s and are used in a variety of commercial and military applications. Examples include DLP projectors, inkjet printers, and accelerometers and MEMS gyroscopes used to deploy automobile airbags.

- Since the early 2000s, the integration of optical functionality (optical computing) into silicon chips has been actively pursued in both academic research and in industry resulting in the successful commercialization of silicon based integrated optical transceivers combining optical devices (modulators, detectors, routing) with CMOS based electronics. Integrated optical circuits are also being developed.

- Integrated circuits are also being developed for sensor applications in medical implants or other bioelectronic devices. Special sealing techniques have to be applied in such biogenic environments to avoid corrosion or biodegradation of the exposed semiconductor materials.

As of 2016, the vast majority of all transistors are fabricated in a single layer on one side of a chip of silicon in a flat 2-dimensional planar process. Researchers have produced prototypes of several promising alternatives, such as:

- various approaches to stacking several layers of transistors to make a three-dimensional integrated circuit, such as through-silicon via, "monolithic 3D", stacked wire bonding, etc.
- transistors built from other materials: graphene transistors, molybdenite transistors, carbon nanotube field-effect transistor, gallium nitride transistor, transistor-like nanowire electronic devices, organic field-effect transistor, etc.
- fabricating transistors over the entire surface of a small sphere of silicon.
- modifications to the substrate, typically to make "flexible transistors" for a flexible display or other flexible electronics, possibly leading to a roll-away computer.

## Design

The cost of designing and developing a complex integrated circuit is quite high, normally in the multiple tens of millions of dollars. This only makes economic sense if production volume is high, so the non-recurring engineering (NRE) costs are spread across typically millions of production units.

Modern semiconductor chips have billions of components, and are too complex to be designed by hand. Software tools to help the designer are essential. Electronic Design Automation (**EDA**), also referred to as Electronic Computer-Aided Design (ECAD), is a category of software tools for designing electronic systems, including integrated circuits. The tools work together in a design flow that engineers use to design and analyze entire semiconductor chips.

## Types

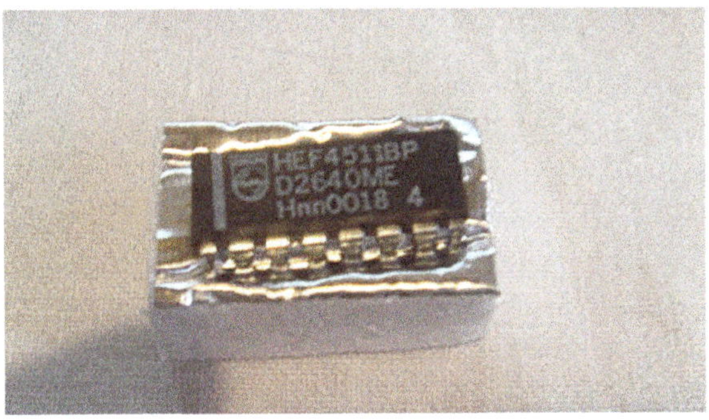


A CMOS 4511 IC in a DIP

Integrated circuits can be classified into analog, digital and mixed signal (both analog and digital on the same chip).

Digital integrated circuits can contain anywhere from one to billions of logic gates, flip-flops, multiplexers, and other circuits in a few square millimeters. The small size of these circuits allows high speed, low power dissipation, and reduced manufacturing cost compared with board-level integration. These digital ICs, typically microprocessors, DSPs, and microcontrollers, work using boolean algebra to process "one" and "zero" signals.

The die from an Intel 8742, an 8-bit microcontroller that includes a CPU running at 12 MHz, 128 bytes of RAM, 2048 bytes of EPROM, and I/O in the same chip

Among the most advanced integrated circuits are the microprocessors or "cores", which control everything from computers and cellular phones to digital microwave ovens. Digital memory chips and application-specific integrated circuits (ASICs) are examples of other families of integrated circuits that are important to the modern information society.

In the 1980s, programmable logic devices were developed. These devices contain circuits whose logical function and connectivity can be programmed by the user, rather than being fixed by the integrated circuit manufacturer. This allows a single chip to be programmed to implement different LSI-type functions such as logic gates, adders and registers. Current devices called field-programmable gate arrays (FPGAs) can (as of 2016) implement the equivalent of millions of gates in parallel and operate up to 1 GHz.

Analog ICs, such as sensors, power management circuits, and operational amplifiers, work by processing continuous signals. They perform functions like amplification, active filtering, demodulation, and mixing. Analog ICs ease the burden on circuit designers by having expertly designed analog circuits available instead of designing a difficult analog circuit from scratch.

ICs can also combine analog and digital circuits on a single chip to create functions such as A/D converters and D/A converters. Such mixed-signal circuits offer smaller size and lower cost, but must carefully account for signal interference. Prior to the late 1990s, radios could not be fabricated in the same low-cost CMOS processes as microprocessors. But since 1998, a large number of radio chips have been developed using CMOS processes. Examples include Intel's DECT cordless phone, or 802.11 (Wi-Fi) chips created by Atheros and other companies.

Modern electronic component distributors often further sub-categorize the huge variety of integrated circuits now available:

- Digital ICs are further sub-categorized as logic ICs, memory chips, interface ICs (level shifters, serializer/deserializer, etc.), Power Management ICs, and programmable devices.

- Analog ICs are further sub-categorized as linear ICs and RF ICs.
- mixed-signal integrated circuits are further sub-categorized as data acquisition ICs (including A/D converters, D/A converter, digital potentiometers) and clock/timing ICs.

## Manufacturing

## Fabrication

Rendering of a small standard cell with three metal layers (dielectric has been removed). The sand-colored structures are metal interconnect, with the vertical pillars being contacts, typically plugs of tungsten. The reddish structures are polysilicon gates, and the solid at the bottom is the crystalline silicon bulk.

The semiconductors of the periodic table of the chemical elements were identified as the most likely materials for a *solid-state vacuum tube*. Starting with copper oxide, proceeding to germanium, then silicon, the materials were systematically studied in the 1940s and 1950s. Today, monocrystalline silicon is the main substrate used for ICs although some III-V compounds of the periodic table such as gallium arsenide are used for specialized applications like LEDs, lasers, solar cells and the highest-speed integrated circuits. It took decades to perfect methods of creating crystals without defects in the crystalline structure of the semiconducting material.

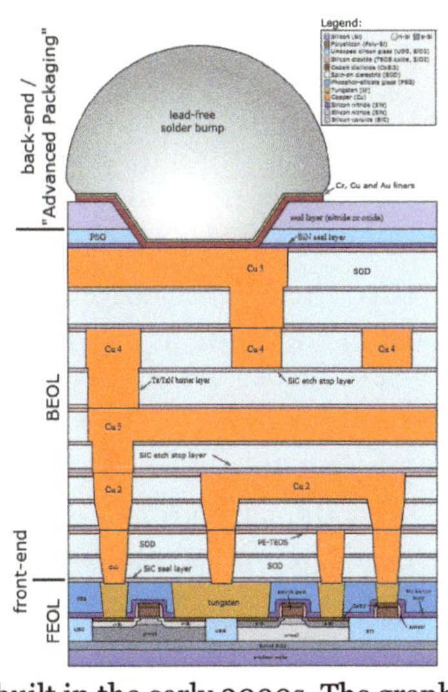


Schematic structure of a CMOS chip, as built in the early 2000s. The graphic shows LDD-MISFET's on an SOI substrate with five metallization layers and solder bump for flip-chip bonding. It also shows the section for FEOL (front-end of line), BEOL (back-end of line) and first parts of back-end process

Semiconductor ICs are fabricated in a planar process which includes three key process steps – imaging, deposition and etching. The main process steps are supplemented by doping and cleaning.

Mono-crystal silicon wafers (or for special applications, silicon on sapphire or gallium arsenide wafers) are used as the *substrate*. Photolithography is used to mark different areas of the substrate to be doped or to have polysilicon, insulators or metal (typically aluminium or copper) tracks deposited on them.

- Integrated circuits are composed of many overlapping layers, each defined by photolithography, and normally shown in different colors. Some layers mark where various dopants are diffused into the substrate (called diffusion layers), some define where additional ions are implanted (implant layers), some define the conductors (polysilicon or metal layers), and some define the connections between the conducting layers (via or contact layers). All components are constructed from a specific combination of these layers.
- In a self-aligned CMOS process, a transistor is formed wherever the gate layer (polysilicon or metal) crosses a diffusion layer.
- Capacitive structures, in form very much like the parallel conducting plates of a traditional electrical capacitor, are formed according to the area of the "plates", with insulating material between the plates. Capacitors of a wide range of sizes are common on ICs.
- Meandering stripes of varying lengths are sometimes used to form on-chip resistors, though most logic circuits do not need any resistors. The ratio of the length of the resistive structure to its width, combined with its sheet resistivity, determines the resistance.
- More rarely, inductive structures can be built as tiny on-chip coils, or simulated by gyrators.

Since a CMOS device only draws current on the *transition* between logic states, CMOS devices consume much less current than bipolar devices.

A random-access memory is the most regular type of integrated circuit; the highest density devices are thus memories; but even a microprocessor will have memory on the chip. Although the structures are intricate – with widths which have been shrinking for decades – the layers remain much thinner than the device widths. The layers of material are fabricated much like a photographic process, although light waves in the visible spectrum cannot be used to "expose" a layer of material, as they would be too large for the features. Thus photons of higher frequencies (typically ultraviolet) are used to create the patterns for each layer. Because each feature is so small, electron microscopes are essential tools for a process engineer who might be debugging a fabrication process.

Each device is tested before packaging using automated test equipment (ATE), in a process known as wafer testing, or wafer probing. The wafer is then cut into rectangular blocks, each of which is called a *die*. Each good die (plural *dice*, *dies*, or *die*) is then connected into a package using aluminium (or gold) bond wires which are thermosonically bonded to *pads*, usually found around the edge of the die. . Thermosonic bonding was first introduced by A. Coucoulas which provided a reliable means of forming these vital electrical connections to the outside world. After packaging, the devices go through final testing on the same or similar ATE used during wafer probing. Industrial CT scanning can also be used. Test cost can account for over 25% of the cost of fabrication on lower-cost products, but can be negligible on low-yielding, larger, or higher-cost devices.

As of 2016, a fabrication facility (commonly known as a *semiconductor fab*) can cost over US$8 billion to construct. The cost of a fabrication facility rises over time (Rock's law) because much of the operation is automated. Today, the most advanced processes employ the following techniques:

- The wafers are up to 300 mm in diameter (wider than a common dinner plate).
- As of 2016, a state of the art foundry can produce 14 nm transistors, as implemented by Intel, TSMC, Samsung, and Global Foundries. The next step, to 10 nm devices, is expected in 2017.
- Copper interconnects where copper wiring replaces aluminium for interconnects.
- Low-K dielectric insulators.
- Silicon on insulator (SOI).
- Strained silicon in a process used by IBM known as strained silicon directly on insulator (SSDOI).
- Multigate devices such as tri-gate transistors being manufactured by Intel from 2011 in their 22 nm process.

## Packaging

A Soviet MSI nMOS chip made in 1977, part of a four-chip calculator set designed in 1970

The earliest integrated circuits were packaged in ceramic flat packs, which continued to be used by the military for their reliability and small size for many years. Commercial circuit packaging quickly moved to the dual in-line package (DIP), first in ceramic and later in plastic. In the 1980s pin counts of VLSI circuits exceeded the practical limit for DIP packaging, leading to pin grid array (PGA) and leadless chip carrier (LCC) packages. Surface mount packaging appeared in the early 1980s and became popular in the late 1980s, using finer lead pitch with leads formed as either gull-wing or J-lead, as exemplified by the small-outline integrated circuit (SOIC) package – a carrier which occupies an area about 30–50% less than an equivalent DIP and is typically 70% thinner. This package has "gull wing" leads protruding from the two long sides and a lead spacing of 0.050 inches.

In the late 1990s, plastic quad flat pack (PQFP) and thin small-outline package (TSOP) packages became the most common for high pin count devices, though PGA packages are still often used for high-end microprocessors. Intel and AMD are currently transitioning from PGA packages on high-end microprocessors to land grid array (LGA) packages.

Ball grid array (BGA) packages have existed since the 1970s. Flip-chip Ball Grid Array packages, which allow for much higher pin count than other package types, were developed in the 1990s. In an FCBGA package the die is mounted upside-down (flipped) and connects to the package balls via a package substrate that is similar to a printed-circuit board rather than by wires. FCBGA packages allow an array of input-output signals (called Area-I/O) to be distributed over the entire die rather than being confined to the die periphery.

Traces going out of the die, through the package, and into the printed circuit board have very different electrical properties, compared to on-chip signals. They require special design techniques and need much more electric power than signals confined to the chip itself.

When multiple dies are put in one package, the result is a System in Package, or SiP. A Multi-Chip Module, or MCM, is created by combining multiple dies on a small substrate often made of ceramic. The distinction between a big MCM and a small printed circuit board is sometimes fuzzy.

## Chip Labeling and Manufacture Date

Most integrated circuits are large enough to include identifying information. Four common sections are the manufacturer's name or logo, the part number, a part production batch number and serial number, and a four-digit date-code to identify when the chip was manufactured. Extremely small surface mount technology parts often bear only a number used in a manufacturer's lookup table to find the chip characteristics.

The manufacturing date is commonly represented as a two-digit year followed by a two-digit week code, such that a part bearing the code 8341 was manufactured in week 41 of 1983, or approximately in October 1983.

## Intellectual Property

The possibility of copying by photographing each layer of an integrated circuit and preparing photomasks for its production on the basis of the photographs obtained is a reason for the introduction of legislation for the protection of layout-designs. The Semiconductor Chip Protection Act of 1984 established intellectual property protection for photomasks used to produce integrated circuits.

A diplomatic conference was held at Washington, D.C., in 1989, which adopted a Treaty on Intellectual Property in Respect of Integrated Circuits (IPIC Treaty).

The Treaty on Intellectual Property in respect of Integrated Circuits, also called Washington Treaty or IPIC Treaty (signed at Washington on 26 May 1989) is currently not in force, but was partially integrated into the TRIPS agreement.

National laws protecting IC layout designs have been adopted in a number of countries, including Japan, the EC, the UK, Australia, and Korea.

## Other Developments

Future developments seem to follow the multi-core multi-microprocessor paradigm, already used by Intel and AMD multi-core processors. Rapport Inc. and IBM started shipping the KC256 in 2006, a 256-core microprocessor. Intel, as recently as February–August 2011, unveiled a prototype, "not for commercial sale" chip that bears 80 cores. Each core is capable of handling its own task independently of the others. This is in response to the heat-versus-speed limit that is about to be reached using existing transistor technology. This design provides a new challenge to chip programming. Parallel programming languages such as the open-source X10 programming language are designed to assist with this task.

## Generations

In the early days of simple integrated circuits, the technology's large scale limited each chip to only a few transistors, and the low degree of integration meant the design process was relatively simple. Manufacturing yields were also quite low by today's standards. As the technology progressed, millions, then billions of transistors could be placed on one chip, and good designs required thorough planning, giving rise to the field of Electronic Design Automation, or EDA.

| Name | Signification | Year | Transistors number | Logic gates number |
|---|---|---|---|---|
| SSI | *small-scale integration* | 1964 | 1 to 10 | 1 to 12 |
| MSI | *medium-scale integration* | 1968 | 10 to 500 | 13 to 99 |
| LSI | *large-scale integration* | 1971 | 500 to 20,000 | 100 to 9,999 |
| VLSI | *very large-scale integration* | 1980 | 20,000 to 1,000,000 | 10,000 to 99,999 |
| ULSI | *ultra-large-scale integration* | 1984 | 1,000,000 and more | 100,000 and more |

## SSI, MSI and LSI

The first integrated circuits contained only a few transistors. Early digital circuits containing tens of transistors provided a few logic gates, and early linear ICs such as the Plessey SL201 or the Philips TAA320 had as few as two transistors. The number of transistors in an integrated circuit has increased dramatically since then. The term "large scale integration" (LSI) was first used by IBM scientist Rolf Landauer when describing the theoretical concept; that term gave rise to the terms "small-scale integration" (SSI), "medium-scale integration" (MSI), "very-large-scale integration" (VLSI), and "ultra-large-scale integration" (ULSI). The early integrated circuits were SSI.

SSI circuits were crucial to early aerospace projects, and aerospace projects helped inspire development of the technology. Both the Minuteman missile and Apollo program needed lightweight digital computers for their inertial guidance systems. Although the Apollo guidance computer led and motivated integrated-circuit technology, it was the Minuteman missile that forced it into mass-production. The Minuteman missile program and various other Navy programs accounted for the total $4 million integrated circuit market in 1962, and by 1968, U.S. Government space and defense spending still accounted for 37% of the $312 million total production.

The demand by the U.S. Government supported the nascent integrated circuit market until costs fell enough to allow IC firms to penetrate first the industrial and eventually the consumer markets. The average price per integrated circuit dropped from $50.00 in 1962 to $2.33 in 1968. Integrated circuits began to appear in consumer products by the turn of the decade, a typical application being FM inter-carrier sound processing in television receivers.

The first MOS chips were small-scale integration chips for NASA satellites.

The next step in the development of integrated circuits, taken in the late 1960s, introduced devices which contained hundreds of transistors on each chip, called "medium-scale integration" (MSI).

In 1964, Frank Wanlass demonstrated a single-chip 16-bit shift register he designed, with an incredible (at the time) 120 transistors on a single chip.

MSI devices were attractive economically because while they cost a little more to produce than SSI devices, they allowed more complex systems to be produced using smaller circuit boards, less assembly work (because of fewer separate components), and a number of other advantages.

Further development, driven by the same economic factors, led to "large-scale integration" (LSI) in the mid-1970s, with tens of thousands of transistors per chip.

The masks used to process and manufacture SSI, MSI and early LSI and VLSI devices (such as the microprocessors of the early 1970s) were mostly created by hand, often using Rubylith-tape or similar. For large or complex ICs (such as memories or processors), this was often done by specially hired layout people under supervision of a team of engineers, who would also, along with the circuit designers, inspect and verify the correctness and completeness of each mask. However, modern VLSI devices contain so many transistors, layers, interconnections, and other features that it is no longer feasible to check the masks or do the original design by hand. The engineer depends on computer programs and other hardware aids to do most of this work.

Integrated circuits such as 1K-bit RAMs, calculator chips, and the first microprocessors, that began to be manufactured in moderate quantities in the early 1970s, had under 4,000 transistors. True LSI circuits, approaching 10,000 transistors, began to be produced around 1974, for computer main memories and second-generation microprocessors.

Some SSI and MSI chips, like discrete transistors, are still mass-produced, both to maintain old equipment and build new devices that require only a few gates. The 7400 series of TTL chips, for example, has become a de facto standard and remains in production.

## VLSI

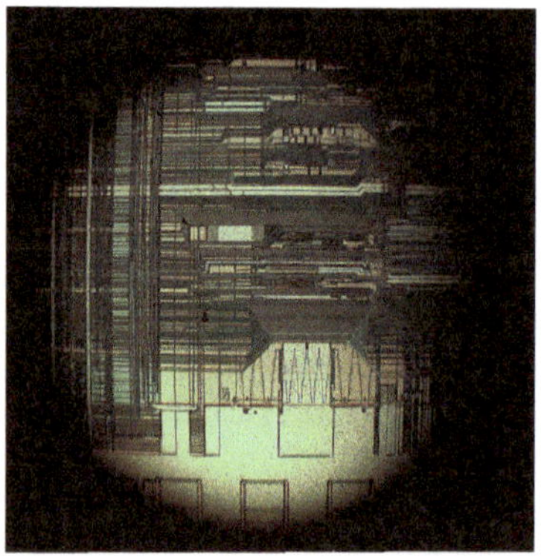

Upper interconnect layers on an Intel 80486DX2 microprocessor die

The final step in the development process, starting in the 1980s and continuing through the present, was "very-large-scale integration" (VLSI). The development started with hundreds of thousands of transistors in the early 1980s, and continues beyond ten billion transistors as of 2016.

Multiple developments were required to achieve this increased density. Manufacturers moved to smaller design rules and cleaner fabrication facilities, so that they could make chips with more transistors and maintain adequate yield. The path of process improvements was summarized by the International Technology Roadmap for Semiconductors (ITRS). Design tools improved enough to make it practical to finish these designs in a reasonable time. The more energy-efficient CMOS replaced NMOS and PMOS, avoiding a prohibitive increase in power consumption.

In 1986 the first one-megabit RAM chips were introduced, containing more than one million transistors. Microprocessor chips passed the million-transistor mark in 1989 and the billion-transistor mark in 2005. The trend continues largely unabated, with chips introduced in 2007 containing tens of billions of memory transistors.

## ULSI, WSI, SOC and 3D-IC

To reflect further growth of the complexity, the term *ULSI* that stands for "ultra-large-scale integration" was proposed for chips of more than 1 million transistors.

Wafer-scale integration (WSI) is a means of building very large integrated circuits that uses an entire silicon wafer to produce a single "super-chip". Through a combination of large size and reduced packaging, WSI could lead to dramatically reduced costs for some systems, notably massively parallel supercomputers. The name is taken from the term Very-Large-Scale Integration, the current state of the art when WSI was being developed.

A system-on-a-chip (SoC or SOC) is an integrated circuit in which all the components needed for a computer or other system are included on a single chip. The design of such a device can be complex and costly, and building disparate components on a single piece of silicon may compromise the efficiency of some elements. However, these drawbacks are offset by lower manufacturing and assembly costs and by a greatly reduced power budget: because signals among the components are kept on-die, much less power is required.

A three-dimensional integrated circuit (3D-IC) has two or more layers of active electronic components that are integrated both vertically and horizontally into a single circuit. Communication between layers uses on-die signaling, so power consumption is much lower than in equivalent separate circuits. Judicious use of short vertical wires can substantially reduce overall wire length for faster operation.

## Silicon Labelling and Graffiti

To allow identification during production most silicon chips will have a serial number in one corner. It is also common to add the manufacturer's logo. Ever since ICs were created, some chip designers have used the silicon surface area for surreptitious, non-functional images or words. These are sometimes referred to as chip art, silicon art, silicon graffiti or silicon doodling.

## ICs and IC Families

- The 555 timer IC

- The 741 operational amplifier
- 7400 series TTL logic building blocks
- 4000 series, the CMOS counterpart to the 7400 series
- Intel 4004, the world's first microprocessor, which led to the famous 8080 CPU and then the IBM PC's 8088, 80286, 486 etc.
- The MOS Technology 6502 and Zilog Z80 microprocessors, used in many home computers of the early 1980s
- The Motorola 6800 series of computer-related chips, leading to the 68000 and 88000 series (used in some Apple computers and in the 1980s Commodore Amiga series).
- The LM-series of analog integrated circuits.

## Basic Semiconductor Physics

- Conductivity in between those of metals and insulators.
- Conductivity can be varied over orders of magnitude by changes in temperature, optical excitation, and impurity content (doping).
- Generally found in column IV and neighboring columns of the periodic table.
- Elemental semiconductors: Si, Ge.
- Compound semiconductors:

Binary :

GaAs, AlAs, GaP, etc. (III-V).

ZnS, ZnTe, CdSe (II-VI).

SiC, SiGe (IV compounds).

Ternary : GaAsP.

Quaternary : InGaAsP.

- Si widely used for rectifiers, transistors, and ICs.
- III-V compounds widely used in optoelectronic and high-speed applications.

### Applications

- Integrated circuits (ICs) SSI, MSI, LSI, and VLSI.
- Fluorescent materials used in TV screens II-VI (ZnS).

- Light detectors InSb, CdSe, PbTe, HgCdTe.
- Infrared and nuclear radiation detectors Si and Ge.
- Gunn diode (microwave device) GaAs, InP.
- Semiconductor LEDs GaAs, GaP.
- Semiconductor LASERs GaAs, AlGaAs.

## Energy Gap

- Distinguishing feature among metals, insulators, and semiconductors.
- Determines the absorption/emission spectra, the leakage current, and the intrinsic conductivity.
- Unique value for each semiconductor (e.g. 1.12 eV for Si, 1.42 eV for GaAs) function of temperature.

## Impurities

- Can be added in precisely controlled amounts.
- Can change the electronic and optical properties.
- Used to vary conductivity over wide ranges.
- Can even change conduction process from conduction by negative charge carriers to positive charge carriers and vice versa.
- Controlled addition of impurities doping.

## Crystal Lattices

- Semiconductor properties can be strongly affected by crystal structure.

## Types of Solids

- Basically, there are three types of solids: crystalline, amorphous, and polycrystalline.

## Crystalline Solids

- Atoms making up the solid arranged in a periodic fashion, repeated throughout.
- Have long-range order.
- Used for IC fabrication.

## Amorphous Solids

- Have no periodic structure at all.

- Interatomic distance and bond angles are almost the same as in the crystalline material of the same substance, however, a long-range order is missing.
- Said to have short-range order.
- a-Si (alloy of amorphous Si with $H_2$ and other similar amorphous alloys) has found important applications in photovoltaic technology and in large-area ICs used in flat displays, printers, copiers, scanners, and imagers.

## Polycrystalline Solids

- Composed of many small regions of single-crystal material of irregular size, separated by grain boundaries.

## Lattice

- 3-D periodic arrangement of atoms in a crystal.
- Defined by primitive basis vectors $\vec{a}, \vec{b}$ *and* $\vec{c}$ , which are three independent shortest vectors connecting lattice sites.
- The coordinates of all points belonging to the crystal lattice are given by vectors $\vec{R}_{k,l,m} = k\vec{a} + l\vec{b} + m\vec{c}$, where k, l, and m are integers.
- Properties of the periodic crystal determine the allowed energies of electrons that participate in the conduction process. Thus, the lattice not only determines the mechanical properties of the crystal, but also its electrical properties.

## Unit Cell

- Representative of the entire crystal and regularly repeated throughout the crystal.
- The crystal can be analyzed as a whole by investigating a representative volume.
- Can find:

  i) the distances between nearest atoms and next nearest atoms,

  ii) the fraction of the unit cell volume filled by atoms, and

  iii) the density of the solid (related to the atomic arrangement).

## Primitive Cell

- Smallest unit cell that can be repeated to form the lattice.

## Cubic Lattices

- Simplest 3-D lattice, where the unit cell is a cube.
  - Three types:

i) simple cubic (sc) (e.g., Ga),

ii) body-centered cubic (bcc) (e.g., Na, W), and

iii) face-centered cubic (fcc) (e.g., Al, Au).

- Lattice constant: the length of each side of the cube.

## The Diamond Lattice

- Two interpenetrating fcc sublattices spaced 1/4th along the body diagonal.
- When the constituent atoms of the two sublattices are different, then the structure is Zincblende (e.g. GaAs).

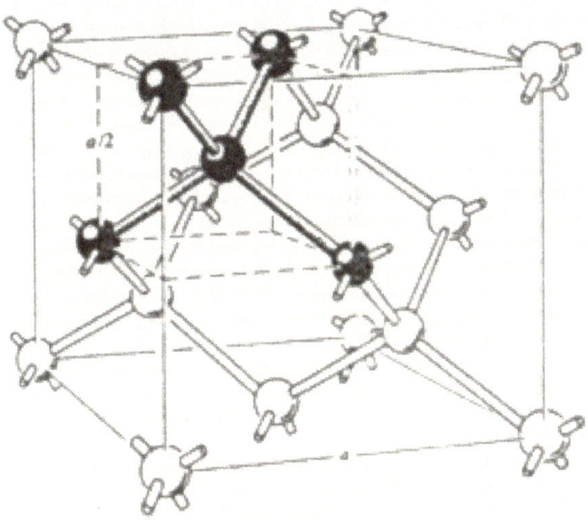

The Diamond lattice.

- Diamond and Zincblende are the two most common crystal structures for cubic semiconductors.
- Each atom in diamond and zincblende lattice is surrounded by four nearest neighbors.
- Tetrahedral configuration.
- By varying the atomic compositions of these two sublattices, one can grow ternary (e.g. $Al_xGa_{1-x}As$) and quaternary (e.g., $In_xGa_{1-x}As_yP_{1-y}$) compounds.

EXAMPLE 1.1: Find the fraction of the unit cell volume filled with hard spheres for a diamond lattice.

SOLUTION: Diamond lattice unit cell consists of an fcc lattice along with 4 inside atoms. The corner atoms in an fcc lattice is shared by seven nearest neighbors, thus each contributing 1/8th of a sphere, and the face atoms are shared by one nearest neighbor, thus each contribution ½ of a sphere.

Thus, Atoms/cell = 1 (corners) + 3 (faces) + 4 (inside) = 8 Nearest neighbor distance = $(a\sqrt{3})/4$

Radius of each sphere = $(a\sqrt{3})/8$ Volume of each sphere = $(4/3)\pi\left[\left(a\sqrt{3}\right)/8\right]^3$

Maximum fraction of cell filled = $\dfrac{no.of\ spheres \times vol.of\ each\ sphere}{total\ vol.of\ cell}$

$$= \frac{(4/3)\pi\left[(a\sqrt{3})/8\right]^3}{a^3} = 0.34$$

Therefore, the unit cell volume filled with hard spheres for a diamond lattice is only 34%, thus, the diamond lattice is relatively loosely packed.

## Planes and Directions

## Miller Indices

- Found by taking the reciprocal of the intercepts of the plane with the coordinate axes, and converting these to integers.
- Define a set of parallel planes.
    - (hkl) a particular plane.
    - {hkl} equivalent (hkl) planes.
    - [hkl] a particular direction, perpendicular to the (hkl) plane for cubic crystals.
    - <hkl> equivalent [hkl] directions.
- Negative intercepts/directions are denoted by placing bars above the integers, e.g. $\bar{h}k\bar{l}$.
- Three main indices (100), (110), and (111).
- Atom densities different for different planes different electronic properties.

EXAMPLE: Show a (643) plane and (643) direction in a cubic lattice.

SOLUTION: (643) plane: reciprocals (1/6,1/4,1/3), converting them to smallest set of integers gives (2,3,4); thus, the plane has (x,y,z) intercepts as (2,3,4).

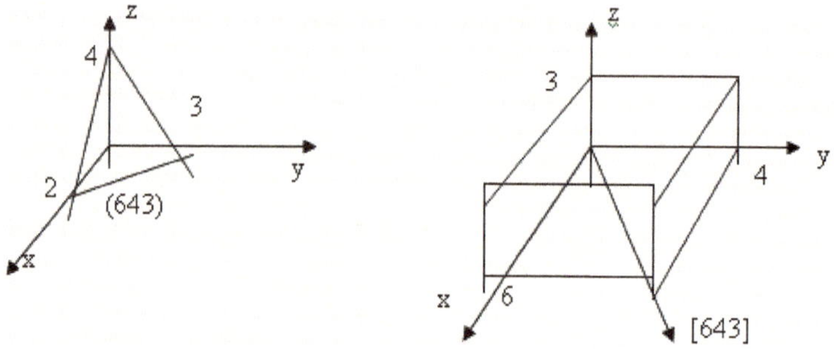

## References


- Lidia Łukasiak & Andrzej Jakubowski (January 2010). "History of Semiconductors" (PDF). Journal of Telecommunication and Information Technology: 3
- Muller, Richard S. & Theodore I. Kamins (1986). Device Electronics for Integrated Circuits. John Wiley and Sons. ISBN 0-471-88758-7
- Lowe, Doug (2013). "Electronics Components: Diodes". Electronics All-In-One Desk Reference For Dummies. John Wiley & Sons. Retrieved January 4, 2013
- Busch, G (1989). "Early history of the physics and chemistry of semiconductors-from doubts to fact in a hundred years". European Journal of Physics. 10 (4): 254–264. Bibcode:1989EJPh...10..254B
- Shockley, William (1950). Electrons and holes in semiconductors : with applications to transistor electronics. R. E. Krieger Pub. Co. ISBN 0-88275-382-7
- "Milestones:Invention of the First Transistor at Bell Telephone Laboratories, Inc., 1947". IEEE Global History Network. IEEE. Retrieved December 7, 2014
- Guthrie, Frederick (October 1873) "On a relation between heat and static electricity," The London, Edinburgh and Dublin Philosophical Magazine and Journal of Science, 4th series, 46 : 257–266
- Louis Nashelsky, Robert L.Boylestad. Electronic Devices and Circuit Theory (9th ed.). India: Prentice-Hall of India Private Limited. pp. 7–10. ISBN 978-81-203-2967-6
- "Der deutsche Erfinder des Transistors – Nachrichten Welt Print – DIE WELT". Welt.de. November 23, 2011. Retrieved May 1, 2016
- Jonscher, A. K. (1961). "The physics of the tunnel diode". British Journal of Applied Physics. 12 (12): 654. Bibcode:1961BJAP...12..654J. doi:10.1088/0508-3443/12/12/304
- Crecraft, David; Stephen Gergely (2002). Analog Electronics: Circuits, Systems and Signal Processing. Butterworth-Heinemann. p. 110. ISBN 0-7506-5095-8
- Narendran, N.; Y. Gu (2005). "Life of LED-based white light sources". IEEE/OSA Journal of Display Technology. 1 (1): 167–171. Bibcode:2005JDisT...1..167N. doi:10.1109/JDT.2005.852510
- Zhong Yuan Chang, Willy M. C. Sansen, Low-Noise Wide-Band Amplifiers in Bipolar and CMOS Technologies, page 31, Springer, 1991 ISBN 0792390962
- Sanders, Robert (June 28, 2005). "Nanofluidic transistor, the basis of future chemical processors". Berkeley.edu. Retrieved June 30, 2012
- Bardsley, J. N. (2004). "International OLED Technology Roadmap". IEEE Journal of Selected Topics in Quantum Electronics. 10: 3–4. doi:10.1109/JSTQE.2004.824077
- Horowitz, Paul; Winfield Hill (1989). The Art of Electronics, 2nd Ed. London: Cambridge University Press. p. 44. ISBN 0-521-37095-7
- "Luminus Products CST-90 Series Datasheet" (PDF). Luminus Devices. Archived from the original (PDF) on 2010-03-31. Retrieved October 25, 2009
- Luginbuhl, C. (2014). "The impact of light source spectral power distribution on sky glow". Journal of Quantitative Spectroscopy and Radiative Transfer. 139: 21–26. doi:10.1016/j.jqsrt.2013.12.004
- Chelikowski, J. (2004) "Introduction: Silicon in all its Forms", p. 1 in Silicon: evolution and future of a technology. P. Siffert and E. F. Krimmel (eds.). Springer, ISBN 3-540-40546-1

# An Overview of Quantum Physics

The branch of physics that studies nature and the behavior of matter and energy is known as quantum physics. It helps in expanding the knowledge on individual behaviors of subatomic particles. Some of the fields where quantum physics is applied are electronics, cryptography and quantum computing. This chapter will provide an integrated understanding of quantum physics.

## Quantum Mechanics

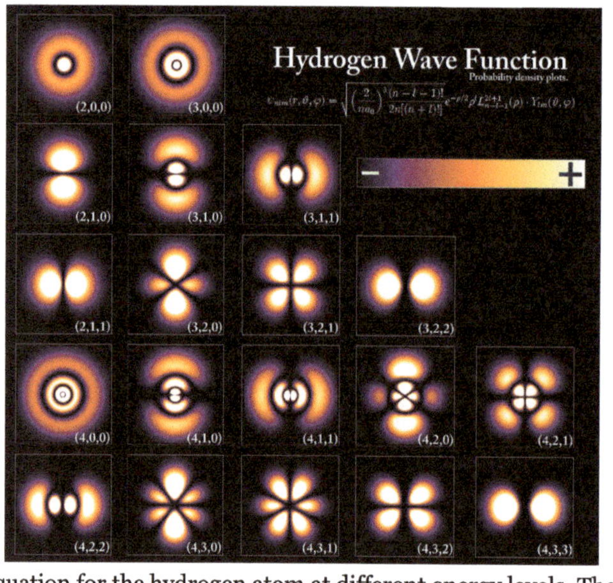


Solution to Schrödinger's equation for the hydrogen atom at different energy levels. The brighter areas represent a higher probability of finding an electron

Quantum mechanics (QM; also known as quantum physics or quantum theory), including quantum field theory, is a branch of physics which is the fundamental theory of nature at small scales and low energy levels of atoms and subatomic particles. Classical physics, the physics existing before quantum mechanics, derives from quantum mechanics as an approximation valid only at large (macroscopic) scales. Quantum mechanics differs from classical physics in that energy, momentum and other quantities are often restricted to discrete values (quantization), objects have characteristics of both particles and waves (wave-particle duality), and there are limits to the precision with which quantities can be known (Uncertainty principle).

Quantum mechanics gradually arose from Max Planck's solution in 1900 to the black-body radiation problem (reported 1859) and Albert Einstein's 1905 paper which offered a quantum-based theory to explain the photoelectric effect (reported 1887). Early quantum theory was profoundly reconceived in the mid-1920s.

The reconceived theory is formulated in various specially developed mathematical formalisms. In one of them, a mathematical function, the wave function, provides information about the probability amplitude of position, momentum, and other physical properties of a particle.

Important applications of quantum theory include quantum chemistry, superconducting magnets, light-emitting diodes, and the laser, the transistor and semiconductors such as the microprocessor, medical and research imaging such as magnetic resonance imaging and electron microscopy, and explanations for many biological and physical phenomena for energy .

## History

Scientific inquiry into the wave nature of light began in the 17th and 18th centuries, when scientists such as Robert Hooke, Christiaan Huygens and Leonhard Euler proposed a wave theory of light based on experimental observations. In 1803, Thomas Young, an English polymath, performed the famous double-slit experiment that he later described in a paper titled *On the nature of light and colours*. This experiment played a major role in the general acceptance of the wave theory of light.

In 1838, Michael Faraday discovered cathode rays. These studies were followed by the 1859 statement of the black-body radiation problem by Gustav Kirchhoff, the 1877 suggestion by Ludwig Boltzmann that the energy states of a physical system can be discrete, and the 1900 quantum hypothesis of Max Planck. Planck's hypothesis that energy is radiated and absorbed in discrete "quanta" (or energy packets) precisely matched the observed patterns of black-body radiation.

In 1896, Wilhelm Wien empirically determined a distribution law of black-body radiation, known as Wien's law in his honor. Ludwig Boltzmann independently arrived at this result by considerations of Maxwell's equations. However, it was valid only at high frequencies and underestimated the radiance at low frequencies. Later, Planck corrected this model using Boltzmann's statistical interpretation of thermodynamics and proposed what is now called Planck's law, which led to the development of quantum mechanics.

Following Max Planck's solution in 1900 to the black-body radiation problem (reported 1859), Albert Einstein offered a quantum-based theory to explain the photoelectric effect (1905, reported 1887). Around 1900-1910, the atomic theory and the corpuscular theory of light first came to be widely accepted as scientific fact; these latter theories can be viewed as quantum theories of matter and electromagnetic radiation, respectively.

Among the first to study quantum phenomena in nature were Arthur Compton, C. V. Raman, and Pieter Zeeman, each of whom has a quantum effect named after him. Robert Andrews Millikan studied the photoelectric effect experimentally, and Albert Einstein developed a theory for it. At the same time, Ernest Rutherford experimentally discovered the nuclear model of the atom, for which Niels Bohr developed his theory of the atomic structure, which was later confirmed by the experiments of Henry Moseley. In 1913, Peter Debye extended Niels Bohr's theory of atomic structure, introducing elliptical orbits, a concept also introduced by Arnold Sommerfeld. This phase is known as old quantum theory.

According to Planck, each energy element ($E$) is proportional to its frequency ($\nu$):

$$E = h\nu,$$

Max Planck is considered the father of the quantum theory

where $h$ is Planck's constant.

Planck cautiously insisted that this was simply an aspect of the *processes* of absorption and emission of radiation and had nothing to do with the *physical reality* of the radiation itself. In fact, he considered his quantum hypothesis a mathematical trick to get the right answer rather than a sizable discovery. However, in 1905 Albert Einstein interpreted Planck's quantum hypothesis realistically and used it to explain the photoelectric effect, in which shining light on certain materials can eject electrons from the material. He won the 1921 Nobel Prize in Physics for this work.

Einstein further developed this idea to show that an electromagnetic wave such as light could also be described as a particle (later called the photon), with a discrete quantum of energy that was dependent on its frequency.

The 1927 Solvay Conference in Brussels

The foundations of quantum mechanics were established during the first half of the 20th century by Max Planck, Niels Bohr, Werner Heisenberg, Louis de Broglie, Arthur Compton, Albert Einstein, Erwin Schrödinger, Max Born, John von Neumann, Paul Dirac, Enrico Fermi, Wolfgang Pauli, Max von Laue, Freeman Dyson, David Hilbert, Wilhelm Wien, Satyendra Nath Bose, Arnold Sommerfeld, and others. The Copenhagen interpretation of Niels Bohr became widely accepted.

In the mid-1920s, developments in quantum mechanics led to its becoming the standard formulation for atomic physics. In the summer of 1925, Bohr and Heisenberg published results that closed the old quantum theory. Out of deference to their particle-like behavior in certain processes and measurements, light quanta came to be called photons (1926). From Einstein's simple postulation was born a flurry of debating, theorizing, and testing. Thus, the entire field of quantum physics emerged, leading to its wider acceptance at the Fifth Solvay Conference in 1927.

It was found that subatomic particles and electromagnetic waves are neither simply particle nor wave but have certain properties of each. This originated the concept of wave–particle duality.

By 1930, quantum mechanics had been further unified and formalized by the work of David Hilbert, Paul Dirac and John von Neumann with greater emphasis on measurement, the statistical nature of our knowledge of reality, and philosophical speculation about the 'observer'. It has since permeated many disciplines including quantum chemistry, quantum electronics, quantum optics, and quantum information science. Its speculative modern developments include string theory and quantum gravity theories. It also provides a useful framework for many features of the modern periodic table of elements, and describes the behaviors of atoms during chemical bonding and the flow of electrons in computer semiconductors, and therefore plays a crucial role in many modern technologies.

While quantum mechanics was constructed to describe the world of the very small, it is also needed to explain some macroscopic phenomena such as superconductors, and superfluids.

The word *quantum* derives from the Latin, meaning "how great" or "how much". In quantum mechanics, it refers to a discrete unit assigned to certain physical quantities such as the energy of an atom at rest. The discovery that particles are discrete packets of energy with wave-like properties led to the branch of physics dealing with atomic and subatomic systems which is today called quantum mechanics. It underlies the mathematical framework of many fields of physics and chemistry, including condensed matter physics, solid-state physics, atomic physics, molecular physics, computational physics, computational chemistry, quantum chemistry, particle physics, nuclear chemistry, and nuclear physics. Some fundamental aspects of the theory are still actively studied.

Quantum mechanics is essential to understanding the behavior of systems at atomic length scales and smaller. If the physical nature of an atom were solely described by classical mechanics, electrons would not *orbit* the nucleus, since orbiting electrons emit radiation (due to circular motion) and would eventually collide with the nucleus due to this loss of energy. This framework was unable to explain the stability of atoms. Instead, electrons remain in an uncertain, non-deterministic, *smeared*, probabilistic wave–particle orbital about the nucleus, defying the traditional assumptions of classical mechanics and electromagnetism.

Quantum mechanics was initially developed to provide a better explanation and description of the atom, especially the differences in the spectra of light emitted by different isotopes of the same chemical element, as well as subatomic particles. In short, the quantum-mechanical atomic model has succeeded spectacularly in the realm where classical mechanics and electromagnetism falter.

Broadly speaking, quantum mechanics incorporates four classes of phenomena for which classical physics cannot account:

- quantization of certain physical properties
- quantum entanglement
- principle of uncertainty
- wave–particle duality

## Mathematical Formulations

In the mathematically rigorous formulation of quantum mechanics developed by Paul Dirac, David Hilbert, John von Neumann, and Hermann Weyl, the possible states of a quantum mechanical system are symbolized as unit vectors (called *state vectors*). Formally, these reside in a complex separable Hilbert space—variously called the *state space* or the *associated Hilbert space* of the system—that is well defined up to a complex number of norm 1 (the phase factor). In other words, the possible states are points in the projective space of a Hilbert space, usually called the complex projective space. The exact nature of this Hilbert space is dependent on the system—for example, the state space for position and momentum states is the space of square-integrable functions, while the state space for the spin of a single proton is just the product of two complex planes. Each observable is represented by a maximally Hermitian (precisely: by a self-adjoint) linear operator acting on the state space. Each eigenstate of an observable corresponds to an eigenvector of the operator, and the associated eigenvalue corresponds to the value of the observable in that eigenstate. If the operator's spectrum is discrete, the observable can attain only those discrete eigenvalues.

In the formalism of quantum mechanics, the state of a system at a given time is described by a complex wave function, also referred to as state vector in a complex vector space. This abstract mathematical object allows for the calculation of probabilities of outcomes of concrete experiments. For example, it allows one to compute the probability of finding an electron in a particular region around the nucleus at a particular time. Contrary to classical mechanics, one can never make simultaneous predictions of conjugate variables, such as position and momentum, to arbitrary precision. For instance, electrons may be considered (to a certain probability) to be located somewhere within a given region of space, but with their exact positions unknown. Contours of constant probability, often referred to as "clouds", may be drawn around the nucleus of an atom to conceptualize where the electron might be located with the most probability. Heisenberg's uncertainty principle quantifies the inability to precisely locate the particle given its conjugate momentum.

According to one interpretation, as the result of a measurement the wave function containing the probability information for a system collapses from a given initial state to a particular eigenstate. The possible results of a measurement are the eigenvalues of the operator representing the observable—which explains the choice of *Hermitian* operators, for which all the eigenvalues are real. The probability distribution of an observable in a given state can be found by computing the spectral decomposition of the corresponding operator. Heisenberg's uncertainty principle is represented by the statement that the operators corresponding to certain observables do not commute.

The probabilistic nature of quantum mechanics thus stems from the act of measurement. This is one of the most difficult aspects of quantum systems to understand. It was the central topic in the famous Bohr–Einstein debates, in which the two scientists attempted to clarify these fundamental principles by way of thought experiments. In the decades after the formulation of quantum mechanics, the question of what constitutes a "measurement" has been extensively studied. Newer interpretations of quantum mechanics have been formulated that do away with the concept of "wave function collapse". The basic idea is that when a quantum system interacts with a measuring apparatus, their respective wave functions become entangled, so that the original quantum system ceases to exist as an independent entity.

Generally, quantum mechanics does not assign definite values. Instead, it makes a prediction using a probability distribution; that is, it describes the probability of obtaining the possible outcomes from measuring an observable. Often these results are skewed by many causes, such as dense probability clouds. Probability clouds are approximate (but better than the Bohr model) whereby electron location is given by a probability function, the wave function eigenvalue, such that the probability is the squared modulus of the complex amplitude, or quantum state nuclear attraction. Naturally, these probabilities will depend on the quantum state at the "instant" of the measurement. Hence, uncertainty is involved in the value. There are, however, certain states that are associated with a definite value of a particular observable. These are known as eigenstates of the observable ("eigen" can be translated from German as meaning "inherent" or "characteristic").

In the everyday world, it is natural and intuitive to think of everything (every observable) as being in an eigenstate. Everything appears to have a definite position, a definite momentum, a definite energy, and a definite time of occurrence. However, quantum mechanics does not pinpoint the exact values of a particle's position and momentum (since they are conjugate pairs) or its energy and time (since they too are conjugate pairs); rather, it provides only a range of probabilities in which that particle might be given its momentum and momentum probability. Therefore, it is helpful to use different words to describe states having *uncertain* values and states having *definite* values (eigenstates). Usually, a system will not be in an eigenstate of the observable (particle) we are interested in. However, if one measures the observable, the wave function will instantaneously be an eigenstate (or "generalized" eigenstate) of that observable. This process is known as wave function collapse, a controversial and much-debated process that involves expanding the system under study to include the measurement device. If one knows the corresponding wave function at the instant before the measurement, one will be able to compute the probability of the wave function collapsing into each of the possible eigenstates. For example, the free particle in the previous example will usually have a wave function that is a wave packet centered around some mean position $x_0$ (neither an eigenstate of position nor of momentum). When one measures the position of the particle, it is impossible to predict with certainty the result. It is probable, but not certain, that it will be near $x_0$, where the amplitude of the wave function is large. After the measurement is performed, having obtained some result $x$, the wave function collapses into a position eigenstate centered at $x$.

The time evolution of a quantum state is described by the Schrödinger equation, in which the Hamiltonian (the operator corresponding to the total energy of the system) generates the time evolution. The time evolution of wave functions is deterministic in the sense that - given a wave function at an *initial* time - it makes a definite prediction of what the wave function will be at any *later* time.

During a measurement, on the other hand, the change of the initial wave function into another, later wave function is not deterministic, it is unpredictable (i.e., random). A time-evolution simulation can be seen here.

Wave functions change as time progresses. The Schrödinger equation describes how wave functions change in time, playing a role similar to Newton's second law in classical mechanics. The Schrödinger equation, applied to the aforementioned example of the free particle, predicts that the center of a wave packet will move through space at a constant velocity (like a classical particle with no forces acting on it). However, the wave packet will also spread out as time progresses, which means that the position becomes more uncertain with time. This also has the effect of turning a position eigenstate (which can be thought of as an infinitely sharp wave packet) into a broadened wave packet that no longer represents a (definite, certain) position eigenstate.

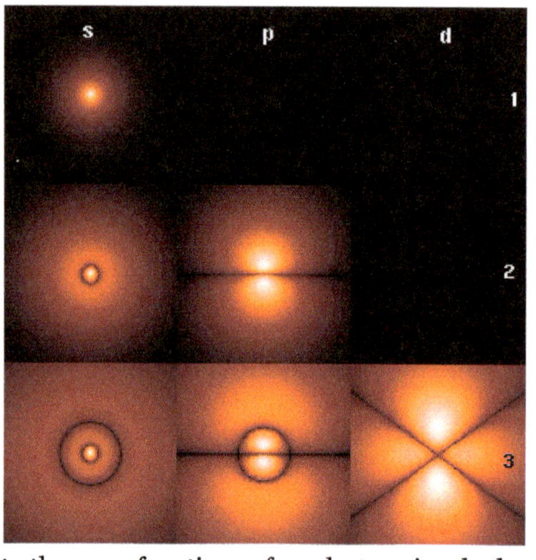


Probability densities corresponding to the wave functions of an electron in a hydrogen atom possessing definite energy levels (increasing from the top of the image to the bottom: $n = 1, 2, 3, \ldots$) and angular momenta (increasing across from left to right: s, p, d, ...). Brighter areas correspond to higher probability density in a position measurement. Such wave functions are directly comparable to Chladni's figures of acoustic modes of vibration in classical physics, and are modes of oscillation as well, possessing a sharp energy and, thus, a definite frequency. The angular momentum and energy are quantized, and take only discrete values like those shown (as is the case for resonant frequencies in acoustics)

Some wave functions produce probability distributions that are constant, or independent of time—such as when in a stationary state of constant energy, time vanishes in the absolute square of the wave function. Many systems that are treated dynamically in classical mechanics are described by such "static" wave functions. For example, a single electron in an unexcited atom is pictured classically as a particle moving in a circular trajectory around the atomic nucleus, whereas in quantum mechanics it is described by a static, spherically symmetric wave function surrounding the nucleus (note, however, that only the lowest angular momentum states, labeled *s*, are spherically symmetric).

The Schrödinger equation acts on the *entire* probability amplitude, not merely its absolute value. Whereas the absolute value of the probability amplitude encodes information about probabilities, its phase encodes information about the interference between quantum states. This gives rise to the "wave-like" behavior of quantum states. As it turns out, analytic solutions of the Schrödinger equation are available for only a very small number of relatively simple model Hamiltonians, of which the quantum harmonic oscillator, the particle in a box, the dihydrogen cation, and the hydrogen atom are the most important representatives. Even the helium atom—which contains just one more electron than does the hydrogen atom—has defied all attempts at a fully analytic treatment.

There exist several techniques for generating approximate solutions, however. In the important method known as perturbation theory, one uses the analytic result for a simple quantum mechanical model to generate a result for a more complicated model that is related to the simpler model by (for one example) the addition of a weak potential energy. Another method is the "semi-classical equation of motion" approach, which applies to systems for which quantum mechanics produces only weak (small) deviations from classical behavior. These deviations can then be computed based on the classical motion. This approach is particularly important in the field of quantum chaos.

## Mathematically Equivalent Formulations of Quantum Mechanics

There are numerous mathematically equivalent formulations of quantum mechanics. One of the oldest and most commonly used formulations is the "transformation theory" proposed by Paul Dirac, which unifies and generalizes the two earliest formulations of quantum mechanics - matrix mechanics (invented by Werner Heisenberg) and wave mechanics (invented by Erwin Schrödinger).

Especially since Werner Heisenberg was awarded the Nobel Prize in Physics in 1932 for the creation of quantum mechanics, the role of Max Born in the development of QM was overlooked until the 1954 Nobel award. The role is noted in a 2005 biography of Born, which recounts his role in the matrix formulation of quantum mechanics, and the use of probability amplitudes. Heisenberg himself acknowledges having learned matrices from Born, as published in a 1940 *festschrift* honoring Max Planck. In the matrix formulation, the instantaneous state of a quantum system encodes the probabilities of its measurable properties, or "observables". Examples of observables include energy, position, momentum, and angular momentum. Observables can be either continuous (e.g., the position of a particle) or discrete (e.g., the energy of an electron bound to a hydrogen atom). An alternative formulation of quantum mechanics is Feynman's path integral formulation, in which a quantum-mechanical amplitude is considered as a sum over all possible classical and non-classical paths between the initial and final states. This is the quantum-mechanical counterpart of the action principle in classical mechanics.

## Interactions with other Scientific Theories

The rules of quantum mechanics are fundamental. They assert that the state space of a system is a Hilbert space and that observables of that system are Hermitian operators acting on that space—although they do not tell us which Hilbert space or which operators. These can be chosen appropriately in order to obtain a quantitative description of a quantum system. An important guide for making these choices is the correspondence principle, which states that the predictions of quantum mechanics reduce to those of classical mechanics when a system moves to higher energies or, equivalently, larger quantum numbers, i.e. whereas a single particle exhibits a degree of randomness, in systems incorporating millions of particles averaging takes over and, at the high energy limit, the statistical probability of random behaviour approaches zero. In other words, classical mechanics is simply a quantum mechanics of large systems. This "high energy" limit is known as the *classical* or *correspondence limit*. One can even start from an established classical model of a particular system, then attempt to guess the underlying quantum model that would give rise to the classical model in the correspondence limit.

When quantum mechanics was originally formulated, it was applied to models whose correspondence limit was non-relativistic classical mechanics. For instance, the well-known model of the quantum harmonic oscillator uses an explicitly non-relativistic expression for the kinetic energy of the oscillator, and is thus a quantum version of the classical harmonic oscillator.

Early attempts to merge quantum mechanics with special relativity involved the replacement of the Schrödinger equation with a covariant equation such as the Klein–Gordon equation or the Dirac equation. While these theories were successful in explaining many experimental results, they had certain unsatisfactory qualities stemming from their neglect of the relativistic creation and annihilation of particles. A fully relativistic quantum theory required the development of quantum field theory, which applies quantization to a field (rather than a fixed set of particles). The first complete quantum field theory, quantum electrodynamics, provides a fully quantum description of the electromagnetic interaction. The full apparatus of quantum field theory is often unnecessary for describing electrodynamic systems. A simpler approach, one that has been employed since the inception of quantum mechanics, is to treat charged particles as quantum mechanical objects being acted on by a classical electromagnetic field. For example, the elementary quantum model of the hydrogen atom describes the electric field of the hydrogen atom using a classical $-e^2 / (4\pi \epsilon_0 r)$ Coulomb potential. This "semi-classical" approach fails if quantum fluctuations in the electromagnetic field play an important role, such as in the emission of photons by charged particles.

Quantum field theories for the strong nuclear force and the weak nuclear force have also been developed. The quantum field theory of the strong nuclear force is called quantum chromodynamics, and describes the interactions of subnuclear particles such as quarks and gluons. The weak nuclear force and the electromagnetic force were unified, in their quantized forms, into a single quantum field theory (known as electroweak theory), by the physicists Abdus Salam, Sheldon Glashow and Steven Weinberg. These three men shared the Nobel Prize in Physics in 1979 for this work.

It has proven difficult to construct quantum models of gravity, the remaining fundamental force. Semi-classical approximations are workable, and have led to predictions such as Hawking radiation. However, the formulation of a complete theory of quantum gravity is hindered by apparent incompatibilities between general relativity (the most accurate theory of gravity currently known) and some of the fundamental assumptions of quantum theory. The resolution of these incompatibilities is an area of active research, and theories such as string theory are among the possible candidates for a future theory of quantum gravity.

Classical mechanics has also been extended into the complex domain, with complex classical mechanics exhibiting behaviors similar to quantum mechanics.

## Quantum Mechanics and Classical Physics

Predictions of quantum mechanics have been verified experimentally to an extremely high degree of accuracy. According to the correspondence principle between classical and quantum mechanics, all objects obey the laws of quantum mechanics, and classical mechanics is just an approximation for large systems of objects (or a statistical quantum mechanics of a large collection of particles). The laws of classical mechanics thus follow from the laws of quantum mechanics as a statistical average at the limit of large systems or large quantum numbers. However, chaotic systems do not have good quantum numbers, and quantum chaos studies the relationship between classical and quantum descriptions in these systems.

Quantum coherence is an essential difference between classical and quantum theories as illustrated by the Einstein–Podolsky–Rosen (EPR) paradox – an attack on a certain philosophical interpretation of quantum mechanics by an appeal to local realism. Quantum interference involves adding together *probability amplitudes*, whereas classical "waves" infer that there is an adding together of *intensities*. For microscopic bodies, the extension of the system is much smaller than the coherence length, which gives rise to long-range entanglement and other nonlocal phenomena characteristic of quantum systems. Quantum coherence is not typically evident at macroscopic scales, though an exception to this rule may occur at extremely low temperatures (i.e. approaching absolute zero) at which quantum behavior may manifest itself macroscopically. This is in accordance with the following observations:

- Many macroscopic properties of a classical system are a direct consequence of the quantum behavior of its parts. For example, the stability of bulk matter (consisting of atoms and molecules which would quickly collapse under electric forces alone), the rigidity of solids, and the mechanical, thermal, chemical, optical and magnetic properties of matter are all results of the interaction of electric charges under the rules of quantum mechanics.
- While the seemingly "exotic" behavior of matter posited by quantum mechanics and relativity theory become more apparent when dealing with particles of extremely small size or velocities approaching the speed of light, the laws of classical, often considered "Newtonian", physics remain accurate in predicting the behavior of the vast majority of "large" objects (on the order of the size of large molecules or bigger) at velocities much smaller than the velocity of light.

## Copenhagen Interpretation of Quantum Versus Classical Kinematics

A big difference between classical and quantum mechanics is that they use very different kinematic descriptions.

In Niels Bohr's mature view, quantum mechanical phenomena are required to be experiments, with complete descriptions of all the devices for the system, preparative, intermediary, and finally measuring. The descriptions are in macroscopic terms, expressed in ordinary language, supplemented with the concepts of classical mechanics. The initial condition and the final condition of the system are respectively described by values in a configuration space, for example a position space, or some equivalent space such as a momentum space. Quantum mechanics does not admit a completely precise description, in terms of both position and momentum, of an initial condition or "state" (in the classical sense of the word) that would support a precisely deterministic and causal prediction of a final condition. In this sense, advocated by Bohr in his mature writings, a quantum phenomenon is a process, a passage from initial to final condition, not an instantaneous "state" in the classical sense of that word. Thus there are two kinds of processes in quantum mechanics: stationary and transitional. For a stationary process, the initial and final condition are the same. For a transition, they are different. Obviously by definition, if only the initial condition is given, the process is not determined. Given its initial condition, prediction of its final condition is possible, causally but only probabilistically, because the Schrödinger equation is deterministic for wave function evolution, but the wave function describes the system only probabilistically.

For many experiments, it is possible to think of the initial and final conditions of the system as being a particle. In some cases it appears that there are potentially several spatially distinct pathways or trajectories by which a particle might pass from initial to final condition. It is an important feature of the quantum kinematic description that it does not permit a unique definite statement of which of those pathways is actually followed. Only the initial and final conditions are definite, and, as stated in the foregoing paragraph, they are defined only as precisely as allowed by the configuration space description or its equivalent. In every case for which a quantum kinematic description is needed, there is always a compelling reason for this restriction of kinematic precision. An example of such a reason is that for a particle to be experimentally found in a definite position, it must be held motionless; for it to be experimentally found to have a definite momentum, it must have free motion; these two are logically incompatible.

Classical kinematics does not primarily demand experimental description of its phenomena. It allows completely precise description of an instantaneous state by a value in phase space, the Cartesian product of configuration and momentum spaces. This description simply assumes or imagines a state as a physically existing entity without concern about its experimental measurability. Such a description of an initial condition, together with Newton's laws of motion, allows a precise deterministic and causal prediction of a final condition, with a definite trajectory of passage. Hamiltonian dynamics can be used for this. Classical kinematics also allows the description of a process analogous to the initial and final condition description used by quantum mechanics. Lagrangian mechanics applies to this. For processes that need account to be taken of actions of a small number of Planck constants, classical kinematics is not adequate; quantum mechanics is needed.

## General Relativity and Quantum Mechanics

Even with the defining postulates of both Einstein's theory of general relativity and quantum theory being indisputably supported by rigorous and repeated empirical evidence, and while they do not directly contradict each other theoretically (at least with regard to their primary claims), they have proven extremely difficult to incorporate into one consistent, cohesive model.

Gravity is negligible in many areas of particle physics, so that unification between general relativity and quantum mechanics is not an urgent issue in those particular applications. However, the lack of a correct theory of quantum gravity is an important issue in physical cosmology and the search by physicists for an elegant "Theory of Everything" (TOE). Consequently, resolving the inconsistencies between both theories has been a major goal of 20th and 21st century physics. Many prominent physicists, including Stephen Hawking, have labored for many years in the attempt to discover a theory underlying *everything*. This TOE would combine not only the different models of subatomic physics, but also derive the four fundamental forces of nature - the strong force, electromagnetism, the weak force, and gravity - from a single force or phenomenon. While Stephen Hawking was initially a believer in the Theory of Everything, after considering Gödel's Incompleteness Theorem, he has concluded that one is not obtainable, and has stated so publicly in his section "Gödel and the End of Physics" (2002).

## Attempts at a Unified Field Theory

The quest to unify the fundamental forces through quantum mechanics is still ongoing. Quantum electrodynamics (or "quantum electromagnetism"), which is currently (in the perturbative regime at least) the most accurately tested physical theory in competition with general relativity, has been successfully merged with the weak nuclear force into the electroweak force and work is currently being done to merge the electroweak and strong force into the electrostrong force. Current predictions state that at around $10^{14}$ GeV the three aforementioned forces are fused into a single unified field. Beyond this "grand unification", it is speculated that it may be possible to merge gravity with the other three gauge symmetries, expected to occur at roughly $10^{19}$ GeV. However — and while special relativity is parsimoniously incorporated into quantum electrodynamics — the expanded general relativity, currently the best theory describing the gravitation force, has not been fully incorporated into quantum theory. One of those searching for a coherent TOE is Edward Witten, a theoretical physicist who formulated the M-theory, which is an attempt at describing the supersymmetrical based string theory. M-theory posits that our apparent 4-dimensional spacetime is, in reality, actually an 11-dimensional spacetime containing 10 spatial dimensions and 1 time dimension, although 7 of the spatial dimensions are - at lower energies - completely "compactified" (or infinitely curved) and not readily amenable to measurement or probing.

Another popular theory is Loop quantum gravity (LQG), a theory first proposed by Carlo Rovelli that describes the quantum properties of gravity. It is also a theory of quantum space and quantum time, because in general relativity the geometry of spacetime is a manifestation of gravity. LQG is an attempt to merge and adapt standard quantum mechanics and standard general relativity. The main output of the theory is a physical picture of space where space is granular. The granularity is a direct consequence of the quantization. It has the same nature of the granularity of the photons in the quantum theory of electromagnetism or the discrete levels of the energy of the atoms. But here it is space itself which is discrete. More precisely, space can be viewed as an extremely fine fabric or network "woven" of finite loops. These networks of loops are called spin networks. The evolution of a spin network over time is called a spin foam. The predicted size of this structure is the Planck length, which is approximately $1.616\times10^{-35}$ m. According to theory, there is no meaning to length shorter than this (cf. Planck scale energy). Therefore, LQG predicts that not just matter, but also space itself, has an atomic structure.

## Philosophical Implications

Since its inception, the many counter-intuitive aspects and results of quantum mechanics have provoked strong philosophical debates and many interpretations. Even fundamental issues, such as Max Born's basic rules concerning probability amplitudes and probability distributions, took decades to be appreciated by society and many leading scientists. Richard Feynman once said, "I think I can safely say that nobody understands quantum mechanics." According to Steven Weinberg, "There is now in my opinion no entirely satisfactory interpretation of quantum mechanics."

The Copenhagen interpretation — due largely to Niels Bohr and Werner Heisenberg — remains most widely accepted amongst physicists, some 75 years after its enunciation. According to this interpretation, the probabilistic nature of quantum mechanics is not a *temporary* feature which will eventually be replaced by a deterministic theory, but instead must be considered a *final* renunciation of the classical idea of "causality." It is also believed therein that any well-defined application of the quantum mechanical formalism must always make reference to the experimental arrangement, due to the conjugate nature of evidence obtained under different experimental situations.

Albert Einstein, himself one of the founders of quantum theory, did not accept some of the more philosophical or metaphysical interpretations of quantum mechanics, such as rejection of determinism and of causality. He is famously quoted as saying, in response to this aspect, "God does not play with dice". He rejected the concept that the state of a physical system depends on the experimental arrangement for its measurement. He held that a state of nature occurs in its own right, regardless of whether or how it might be observed. In that view, he is supported by the currently accepted definition of a quantum state, which remains invariant under arbitrary choice of configuration space for its representation, that is to say, manner of observation. He also held that underlying quantum mechanics there should be a theory that thoroughly and directly expresses the rule against action at a distance; in other words, he insisted on the principle of locality. He considered, but rejected on theoretical grounds, a particular proposal for hidden variables to obviate the indeterminism or acausality of quantum mechanical measurement. He considered that quantum mechanics was a currently valid but not a permanently definitive theory for quantum phenomena. He thought its future replacement would require profound conceptual advances, and would not come quickly or easily. The Bohr-Einstein debates provide a vibrant critique of the Copenhagen Interpretation from an epistemological point of view. In arguing for his views, he produced a series of objections, the most famous of which has become known as the Einstein–Podolsky–Rosen paradox.

John Bell showed that this "EPR" paradox led to experimentally testable differences between quantum mechanics and theories that rely on added hidden variables. Experiments have been performed confirming the accuracy of quantum mechanics, thereby demonstrating that quantum mechanics cannot be improved upon by addition of hidden variables. Alain Aspect's initial experiments in 1982, and many subsequent experiments since, have definitively verified quantum entanglement.

Entanglement, as demonstrated in Bell-type experiments, does not, however, violate causality, since no transfer of information happens. Quantum entanglement forms the basis of quantum cryptography, which is proposed for use in high-security commercial applications in banking and government.

The Everett many-worlds interpretation, formulated in 1956, holds that *all* the possibilities described by quantum theory *simultaneously* occur in a multiverse composed of mostly independent parallel universes. This is not accomplished by introducing some "new axiom" to quantum mechanics, but on the contrary, by *removing* the axiom of the collapse of the wave packet. *All* of the possible consistent states of the measured system and the measuring apparatus (including the observer) are present in a *real* physical - not just formally mathematical, as in other interpretations - quantum superposition. Such a superposition of consistent state combinations of different systems is called an entangled state. While the multiverse is deterministic, we perceive non-deterministic behavior governed by probabilities, because we can only observe the universe (i.e., the consistent state contribution to the aforementioned superposition) that we, as observers, inhabit. Everett's interpretation is perfectly consistent with John Bell's experiments and makes them intuitively understandable. However, according to the theory of quantum decoherence, these "parallel universes" will never be accessible to us. The inaccessibility can be understood as follows: once a measurement is done, the measured system becomes entangled with *both* the physicist who measured it *and* a huge number of other particles, some of which are photons flying away at the speed of light towards the other end of the universe. In order to prove that the wave function did not collapse, one would have to bring *all* these particles back and measure them again, together with the system that was originally measured. Not only is this completely impractical, but even if one *could* theoretically do this, it would have to destroy any evidence that the original measurement took place (including the physicist's memory). In light of these Bell tests, Cramer (1986) formulated his transactional interpretation. Relational quantum mechanics appeared in the late 1990s as the modern derivative of the Copenhagen Interpretation.

## Applications

Quantum mechanics has had enormous success in explaining many of the features of our universe. Quantum mechanics is often the only tool available that can reveal the individual behaviors of the subatomic particles that make up all forms of matter (electrons, protons, neutrons, photons, and others). Quantum mechanics has strongly influenced string theories, candidates for a Theory of Everything.

Quantum mechanics is also critically important for understanding how individual atoms combine covalently to form molecules. The application of quantum mechanics to chemistry is known as quantum chemistry. Relativistic quantum mechanics can, in principle, mathematically describe most of chemistry. Quantum mechanics can also provide quantitative insight into ionic and covalent bonding processes by explicitly showing which molecules are energetically favorable to which others and the magnitudes of the energies involved. Furthermore, most of the calculations performed in modern computational chemistry rely on quantum mechanics.

In many aspects modern technology operates at a scale where quantum effects are significant.

## Electronics

Many modern electronic devices are designed using quantum mechanics. Examples include the laser, the transistor (and thus the microchip), the electron microscope, and magnetic resonance imaging (MRI). The study of semiconductors led to the invention of the diode and the transistor, which are indispensable parts of modern electronics systems, computer and telecommunication devices. Another application is the light emitting diode which is a high-efficiency source of light.

A working mechanism of a resonant tunneling diode device, based on the phenomenon of quantum tunneling through potential barriers. (Left: band diagram; Center: transmission coefficient; Right: current-voltage characteristics) As shown in the band diagram(left), although there are two barriers, electrons still tunnel through via the confined states between two barriers(center), conducting current.

Many electronic devices operate under effect of quantum tunneling. It even exists in the simple light switch. The switch would not work if electrons could not quantum tunnel through the layer of oxidation on the metal contact surfaces. Flash memory chips found in USB drives use quantum tunneling to erase their memory cells. Some negative differential resistance devices also utilize quantum tunneling effect, such as resonant tunneling diode. Unlike classical diodes, its current is carried by resonant tunneling through two potential barriers. Its negative resistance behavior can only be understood with quantum mechanics: As the confined state moves close to Fermi level, tunnel current increases. As it moves away, current decreases. Quantum mechanics is vital to understanding and designing such electronic devices.

## Cryptography

Researchers are currently seeking robust methods of directly manipulating quantum states. Efforts are being made to more fully develop quantum cryptography, which will theoretically allow guaranteed secure transmission of information.

## Quantum Computing

A more distant goal is the development of quantum computers, which are expected to perform certain computational tasks exponentially faster than classical computers. Instead of using classical bits, quantum computers use qubits, which can be in superpositions of states. Another active research topic is quantum teleportation, which deals with techniques to transmit quantum information over arbitrary distances.

## Macroscale Quantum Effects

While quantum mechanics primarily applies to the smaller atomic regimes of matter and energy, some systems exhibit quantum mechanical effects on a large scale. Superfluidity, the frictionless flow of a liquid at temperatures near absolute zero, is one well-known example. So is the closely related phenomenon of superconductivity, the frictionless flow of an electron gas in a conducting material (an electric current) at sufficiently low temperatures. The fractional quantum hall effect is a topological ordered state which corresponds to patterns of long-range quantum entanglement. States with different topological orders (or different patterns of long range entanglements) cannot change into each other without a phase transition.

## Quantum Theory

Quantum theory also provides accurate descriptions for many previously unexplained phenomena, such as black-body radiation and the stability of the orbitals of electrons in atoms. It has also given insight into the workings of many different biological systems, including smell receptors and protein structures. Recent work on photosynthesis has provided evidence that quantum correlations play an essential role in this fundamental process of plants and many other organisms. Even so, classical physics can often provide good approximations to results otherwise obtained by quantum physics, typically in circumstances with large numbers of particles or large quantum numbers. Since classical formulas are much simpler and easier to compute than quantum formulas, classical approximations are used and preferred when the system is large enough to render the effects of quantum mechanics insignificant.

## Examples

## Free Particle

For example, consider a free particle. In quantum mechanics, there is wave–particle duality, so the properties of the particle can be described as the properties of a wave. Therefore, its quantum state can be represented as a wave of arbitrary shape and extending over space as a wave function. The position and momentum of the particle are observables. The Uncertainty Principle states that both the position and the momentum cannot simultaneously be measured with complete precision. However, one *can* measure the position (alone) of a moving free particle, creating an eigenstate of position with a wave function that is very large (a Dirac delta) at a particular position $x$, and zero everywhere else. If one performs a position measurement on such a wave function, the resultant $x$ will be obtained with 100% probability (i.e., with full certainty, or complete precision). This is called an eigenstate of position—or, stated in mathematical terms, a *generalized position eigenstate (eigendistribution)*. If the particle is in an eigenstate of position, then its momentum is completely unknown. On the other hand, if the particle is in an eigenstate of momentum, then its position is completely unknown. In an eigenstate of momentum having a plane wave form, it can be shown that the wavelength is equal to $h/p$, where $h$ is Planck's constant and $p$ is the momentum of the eigenstate.

## Step Potential

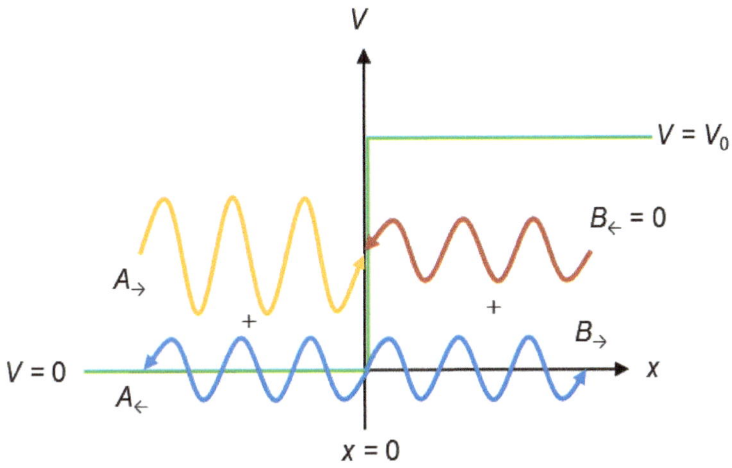


Scattering at a finite potential step of height $V_0$, shown in green. The amplitudes and direction of left- and right-moving waves are indicated. Yellow is the incident wave, blue are reflected and transmitted waves, red does not occur. $E > V_0$ for this figure

The potential in this case is given by:

$$V(x) = \begin{cases} 0, & x < 0, \\ V_0, & x \geq 0. \end{cases}$$

The solutions are superpositions of left- and right-moving waves:

$$\psi_1(x) = \frac{1}{\sqrt{k_1}}\left(A_{\rightarrow}e^{ik_1x} + A_{\leftarrow}e^{-ik_1x}\right) \quad x < 0$$

$$\psi_2(x) = \frac{1}{\sqrt{k_2}}\left(B_{\rightarrow}e^{ik_2x} + B_{\leftarrow}e^{-ik_2x}\right) \quad x > 0$$

where the wave vectors are related to the energy via

$k_1 = \sqrt{2mE/\hbar^2}$, and

$$k_2 = \sqrt{2m(E - V_0)/\hbar^2}$$

with coefficients A and B determined from the boundary conditions and by imposing a continuous derivative on the solution.

Each term of the solution can be interpreted as an incident, reflected, or transmitted component of the wave, allowing the calculation of transmission and reflection coefficients. Notably, in contrast to classical mechanics, incident particles with energies greater than the potential step are partially reflected.

## Rectangular Potential Barrier

This is a model for the quantum tunneling effect which plays an important role in the performance of modern technologies such as flash memory and scanning tunneling microscopy. Quantum tunneling is central to physical phenomena involved in superlattices.

## Particle in a Box

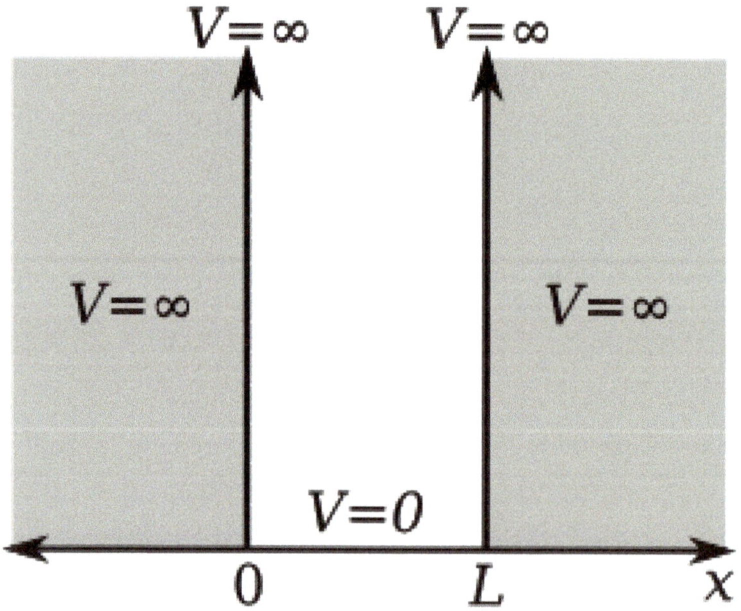

1-dimensional potential energy box (or infinite potential well)

The particle in a one-dimensional potential energy box is the most mathematically simple example where restraints lead to the quantization of energy levels. The box is defined as having zero potential energy everywhere *inside* a certain region, and infinite potential energy everywhere *outside* that region. For the one-dimensional case in the $x$ direction, the time-independent Schrödinger equation may be written

$$-\frac{\hbar^2}{2m}\frac{d^2\psi}{dx^2} = E\psi.$$

With the differential operator defined by

$$\hat{p}_x = -i\hbar \frac{d}{dx}$$

the previous equation is evocative of the classic kinetic energy analogue,

$$\frac{1}{2m}\hat{p}_x^2 = E,$$

with state $\psi$ in this case having energy $E$ coincident with the kinetic energy of the particle.

The general solutions of the Schrödinger equation for the particle in a box are

$$\psi(x) = Ae^{ikx} + Be^{-ikx} \qquad\qquad E = \frac{\hbar^2 k^2}{2m}$$

or, from Euler's formula,

$$\psi(x) = C \sin kx + D \cos kx.$$

The infinite potential walls of the box determine the values of $C$, $D$, and $k$ at $x = 0$ and $x = L$ where $\psi$ must be zero. Thus, at $x = 0$,

$$\psi(0) = 0 = C \sin 0 + D \cos 0 = D$$

and $D = 0$. At $x = L$,

$$\psi(L) = 0 = C \sin kL.$$

in which $C$ cannot be zero as this would conflict with the Born interpretation. Therefore, since $\sin(kL) = 0$, $kL$ must be an integer multiple of $\pi$,

$$k = \frac{n\pi}{L} \qquad\qquad n = 1, 2, 3, \ldots.$$

The quantization of energy levels follows from this constraint on $k$, since

$$E = \frac{\hbar^2 \pi^2 n^2}{2mL^2} = \frac{n^2 h^2}{8mL^2}.$$

## Finite Potential Well

A finite potential well is the generalization of the infinite potential well problem to potential wells having finite depth.

The finite potential well problem is mathematically more complicated than the infinite particle-in-a-box problem as the wave function is not pinned to zero at the walls of the well. Instead, the wave function must satisfy more complicated mathematical boundary conditions as it is nonzero in regions outside the well.

## Harmonic Oscillator

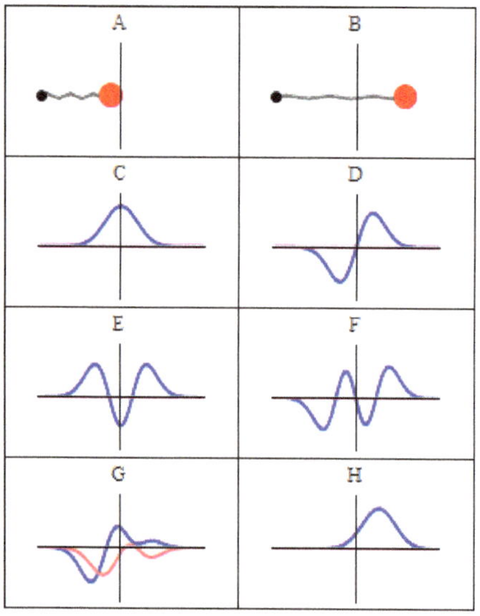


Some trajectories of a harmonic oscillator (i.e. a ball attached to a spring) in classical mechanics (A-B) and quantum mechanics (C-H). In quantum mechanics, the position of the ball is represented by a wave (called the wave function), with the real part shown in blue and the imaginary part shown in red. Some of the trajectories (such as C,D,E,and F) are standing waves (or "stationary states"). Each standing-wave frequency is proportional to a possible energy level of the oscillator. This "energy quantization" does not occur in classical physics, where the oscillator can have *any* energy

As in the classical case, the potential for the quantum harmonic oscillator is given by

$$V(x) = \frac{1}{2} m\omega^2 x^2.$$

This problem can either be treated by directly solving the Schrödinger equation, which is not trivial, or by using the more elegant "ladder method" first proposed by Paul Dirac. The eigenstates are given by

$$\psi_n(x) = \sqrt{\frac{1}{2^n n!}} \cdot \left(\frac{m\omega}{\pi\hbar}\right)^{1/4} \cdot e^{-\frac{m\omega x^2}{2\hbar}} \cdot H_n\left(\sqrt{\frac{m\omega}{\hbar}} x\right),$$

$$n = 0, 1, 2, \ldots.$$

where $H_n$ are the Hermite polynomials

$$H_n(x) = (-1)^n e^{x^2} \frac{d^n}{dx^n}\left(e^{-x^2}\right),$$

and the corresponding energy levels are

$$E_n = \hbar\omega\left(n + \frac{1}{2}\right).$$

This is another example illustrating the quantification of energy for bound states.

# Schrödinger Equation

Schrödinger equation as part of a monument in front of Warsaw University's Centre of New Technologies

In quantum mechanics, the Schrödinger equation is a mathematical equation that describes the evolution over time of a physical system in which quantum effects, such as wave–particle duality, are significant. The equation is a mathematical formulation for studying quantum mechanical systems. It is considered a central result in the study of quantum system and its derivation was a significant landmark in developing the theory of quantum mechanics. It was named after Erwin Schrödinger, who derived the equation in 1925 and published it in 1926, forming the basis for his work that resulted in Schrödinger being awarded the Nobel Prize in Physics in 1933. The equation is a type of differential equation known as a wave-equation, which serves as a mathematical model of the movement of waves.

In classical mechanics, Newton's second law (F = *m*a) is used to make a mathematical prediction as to what path a given system will take following a set of known initial conditions. In quantum mechanics, the analogue of Newton's law is Schrödinger's equation for a quantum system (usually atoms, molecules, and subatomic particles whether free, bound, or localised). It is not a simple algebraic equation, but in general a linear partial differential equation, describing the time-evolution of the system's wave function (also called a "state function").

The concept of a wavefunction is a fundamental postulate of quantum mechanics. Using these postulates, Schrödinger's equation can be derived from the fact that the time-evolution operator must be unitary and must therefore be generated by the exponential of a self-adjoint operator, which is the quantum Hamiltonian. This derivation is explained below.

In the Copenhagen interpretation of quantum mechanics, the wave function is the most complete description that can be given of a physical system. Solutions to Schrödinger's equation describe not only molecular, atomic, and subatomic systems, but also macroscopic systems, possibly even the whole universe. Schrödinger's equation is central to all applications of quantum mechanics including quantum field theory which combines special relativity with quantum mechanics. Theories of quantum gravity, such as string theory, also do not modify Schrödinger's equation.

The Schrödinger equation is not the only way to study quantum mechanical systems and make predictions, as there are other quantum mechanical formulations such as matrix mechanics, introduced by Werner Heisenberg, and path integral formulation, developed chiefly by Richard Feynman. Paul Dirac incorporated matrix mechanics and the Schrödinger equation into a single formulation.

## Equation

### Time-dependent Equation

The form of the Schrödinger equation depends on the physical situation. The most general form is the time-dependent Schrödinger equation, which gives a description of a system evolving with time:

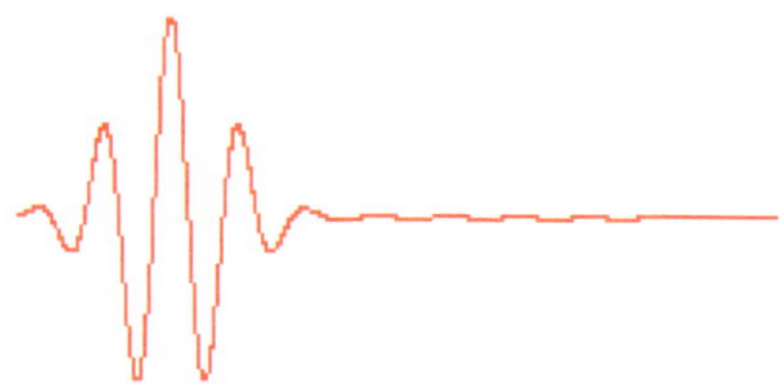

A wave function that satisfies the nonrelativistic Schrödinger equation with $V = 0$. In other words, this corresponds to a particle traveling freely through empty space. The real part of the wave function is plotted here

Time-dependent Schrödinger equation *(general)*

$$i\hbar\frac{\partial}{\partial t}\Psi(\mathbf{r},t) = \hat{H}\Psi(\mathbf{r},t)$$

where $i$ is the imaginary unit, $\hbar$ is the reduced Planck constant which is : $\hbar = \frac{h}{2\pi}$, the symbol $\partial/\partial t$ indicates a partial derivative with respect to time $t$, $\Psi$ is the wave function of the quantum system, r and $t$ are the position vector and time respectively, and $\hat{H}$ is the Hamiltonian operator (which characterizes the total energy of any given wave function and takes different forms depending on the situation).

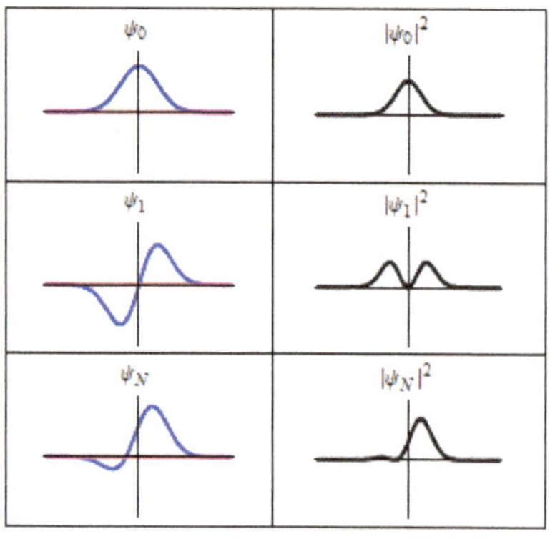


Each of these three rows is a wave function which satisfies the time-dependent Schrödinger equation for a harmonic oscillator. Left: The real part (blue) and imaginary part (red) of the wave function. Right: The probability distribution of finding the particle with this wave function at a given position. The top two rows are examples of stationary states, which correspond to standing waves. The bottom row is an example of a state which is *not* a stationary state. The right column illustrates why stationary states are called "stationary"

The most famous example is the nonrelativistic Schrödinger equation for a single particle moving in an electric field:

Time-dependent Schrödinger equation *(single nonrelativistic particle)*

$$i\hbar\frac{\partial}{\partial t}\Psi(\mathbf{r},t)=\left[\frac{-\hbar^2}{2\mu}\nabla^2+V(\mathbf{r},t)\right]\Psi(\mathbf{r},t)$$

where $\mu$ is the particle's "reduced mass", $V$ is its potential energy, $\nabla^2$ is the Laplacian (a differential operator), and $\Psi$ is the wave function (more precisely, in this context, it is called the "position-space wave function"). In plain language, it means "total energy equals kinetic energy plus potential energy", but the terms take unfamiliar forms for reasons explained below.

Given the particular differential operators involved, this is a linear partial differential equation. It is also a diffusion equation, but unlike the heat equation, this one is also a wave equation given the imaginary unit present in the transient term.

The term *"Schrödinger equation"* can refer to both the general equation (first box above), or the specific nonrelativistic version (second box above and variations thereof). The general equation is indeed quite general, used throughout quantum mechanics, for everything from the Dirac equation to quantum field theory, by plugging in various complicated expressions for the Hamiltonian. The specific nonrelativistic version is a simplified approximation to reality, which is quite accurate in many situations, but very inaccurate in others.

To apply the Schrödinger equation, the Hamiltonian operator is set up for the system, accounting for the kinetic and potential energy of the particles constituting the system, then inserted into the Schrödinger equation. The resulting partial differential equation is solved for the wave function, which contains information about the system.

## Time-independent Equation

The time-dependent Schrödinger equation described above predicts that wave functions can form standing waves, called stationary states (also called "orbitals", as in atomic orbitals or molecular orbitals). These states are important in their own right, and if the stationary states are classified and understood, then it becomes easier to solve the time-dependent Schrödinger equation for *any* state. Stationary states can also be described by a simpler form of the Schrödinger equation, the *time-independent Schrödinger equation.* (This is only used when the Hamiltonian itself is not dependent on time explicitly. However, even in this case the total wave function still has a time dependency.)

Time-independent Schrödinger equation (*general*)

$$\hat{\mathrm{H}}\Psi=E\Psi$$

In words, the equation states:

> *When the Hamiltonian operator acts on a certain wave function $\Psi$, and the result is proportional to the same wave function $\Psi$, then $\Psi$ is a stationary state, and the proportionality constant, E, is the energy of the state $\Psi$.*

The time-independent Schrödinger equation is discussed further below. In linear algebra terminology, this equation is an eigenvalue equation.

As before, the most famous manifestation is the nonrelativistic Schrödinger equation for a single particle moving in an electric field (but not a magnetic field):

Time-independent Schrödinger equation (*single nonrelativistic particle*)

$$\left[\frac{-\hbar^2}{2\mu}\nabla^2 + V(\mathbf{r})\right]\Psi(\mathbf{r}) = E\Psi(\mathbf{r})$$

with definitions as above.

## Derivation

In the modern understanding of quantum mechanics, Schrödinger's equation may be derived as follows. If the wave-function at time t is given by $\Psi(t)$, then by the linearity of quantum mechanics the wave-function at time t' must be given by $\Psi(t') = U(t',t)\Psi(t)$, where $U(t',t)$ is a linear operator. Since time-evolution must preserve the norm of the wave-function, it follows that $U(t',t)$ must be a member of the unitary group of operators acting on wave-functions. We also know that when $t' = t$, we must have $U(t,t) = 1$. Therefore, expanding the operator $U(t',t)$ for t' close to t, we can write $U(t',t) = 1 - iH(t'-t)$ where H is a Hermitian operator. This follows from the fact that the Lie algebra corresponding to the unitary group comprises Hermitian operators. Taking the limit as the time-difference $t'-t$ becomes very small, we obtain Schrödinger's equation.

So far, H is only an abstract Hermitian operator. However using the correspondence principle it is possible to show that, in the classical limit, the expectation value of H is indeed the classical energy. The correspondence principle does not completely fix the form of the quantum Hamiltonian due to the uncertainty principle and therefore the precise form of the quantum Hamiltonian must be fixed empirically.

## Implications

The Schrödinger equation and its solutions introduced a breakthrough in thinking about physics. Schrödinger's equation was the first of its type, and solutions led to consequences that were very unusual and unexpected for the time.

## Total, Kinetic, and Potential Energy

The *overall* form of the equation is *not* unusual or unexpected, as it uses the principle of the conservation of energy. The terms of the nonrelativistic Schrödinger equation can be interpreted as total energy of the system, equal to the system kinetic energy plus the system potential energy. In this respect, it is just the same as in classical physics.

## Quantization

The Schrödinger equation predicts that if certain properties of a system are measured, the result may be *quantized*, meaning that only specific discrete values can occur. One example is *energy quantization*: the energy of an electron in an atom is always one of the quantized energy levels, a fact discovered via atomic spectroscopy. (Energy quantization is discussed below.) Another example is quantization of angular momentum. This was an *assumption* in the earlier Bohr model of the atom, but it is a *prediction* of the Schrödinger equation.

Another result of the Schrödinger equation is that not every measurement gives a quantized result in quantum mechanics. For example, position, momentum, time, and (in some situations) energy can have any value across a continuous range.

## Measurement and Uncertainty

In classical mechanics, a particle has, at every moment, an exact position and an exact momentum. These values change deterministically as the particle moves according to Newton's laws. Under the Copenhagen interpretation of quantum mechanics, particles do not have exactly determined properties, and when they are measured, the result is randomly drawn from a probability distribution. The Schrödinger equation predicts what the probability distributions are, but fundamentally cannot predict the exact result of each measurement.

The Heisenberg uncertainty principle is the statement of the inherent measurement uncertainty in quantum mechanics. It states that the more precisely a particle's position is known, the less precisely its momentum is known, and vice versa.

The Schrödinger equation describes the (deterministic) evolution of the wave function of a particle. However, even if the wave function is known exactly, the result of a specific measurement on the wave function is uncertain.

## Quantum Tunneling

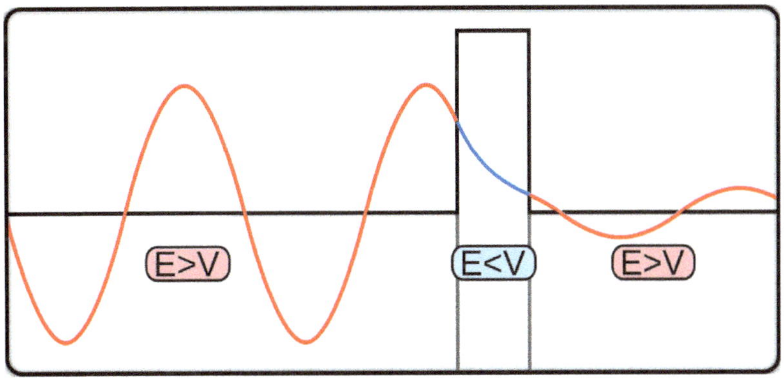


Quantum tunneling through a barrier. A particle coming from the left does not have enough energy to climb the barrier. However, it can sometimes "tunnel" to the other side

In classical physics, when a ball is rolled slowly up a large hill, it will come to a stop and roll back, because it doesn't have enough energy to get over the top of the hill to the other side. However, the Schrödinger equation predicts that there is a small probability that the ball will get to the other side of the hill, even if it has too little energy to reach the top. This is called quantum tunneling. It is related to the distribution of energy: although the ball's assumed position seems to be on one side of the hill, there is a chance of finding it on the other side.

## Particles as Waves

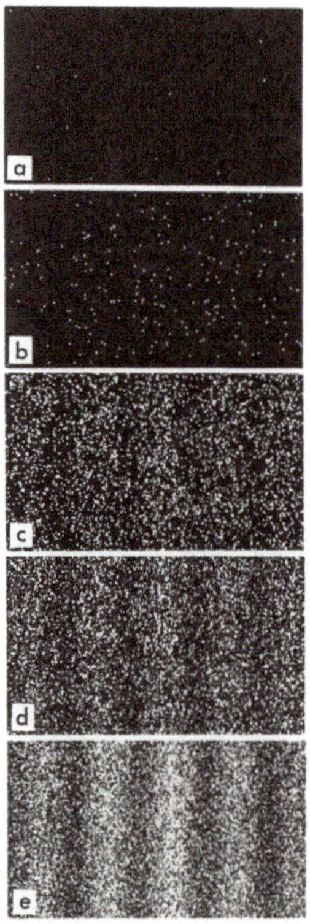


A double slit experiment showing the accumulation of electrons on a screen as time passes

The nonrelativistic Schrödinger equation is a type of partial differential equation called a wave equation. Therefore, it is often said particles can exhibit behavior usually attributed to waves. In some modern interpretations this description is reversed – the quantum state, i.e. wave, is the only genuine physical reality, and under the appropriate conditions it can show features of particle-like behavior. However, Ballentine shows that such an interpretation has problems. Ballentine points out that whilst it is arguable to associate a physical wave with a single particle, there is still only *one* Schrödinger wave equation for many particles. He points out:

> *"If a physical wave field were associated with a particle, or if a particle were identified with a wave packet, then corresponding to N interacting particles there should be N interacting waves in ordinary three-dimensional space. But according to (4.6) that is not the case; instead there is one "wave" function in an abstract 3N-dimensional configuration space. The misinterpretation of psi as a physical wave in ordinary space is possible only because the most common applications of quantum mechanics are to one-particle states, for which configuration space and ordinary space are isomorphic."*

Two-slit diffraction is a famous example of the strange behaviors that waves regularly display, that are not intuitively associated with particles. The overlapping waves from the two slits cancel each other out in some locations, and reinforce each other in other locations, causing a complex pattern to emerge. Intuitively, one would not expect this pattern from firing a single particle at the slits, because the particle should pass through one slit or the other, not a complex overlap of both.

However, since the Schrödinger equation is a wave equation, a single particle fired through a double-slit *does* show this same pattern. Note: The experiment must be repeated many times for the complex pattern to emerge. Although this is counterintuitive, the prediction is correct; in particular, electron diffraction and neutron diffraction are well understood and widely used in science and engineering.

Related to diffraction, particles also display superposition and interference.

The superposition property allows the particle to be in a quantum superposition of two or more quantum states at the same time. However, it is noted that a "quantum state" in quantum mechanics means the *probability* that a system will be, for example at a position $x$, not that the system will actually be at position $x$. It does not imply that the particle itself may be in two classical states at once. Indeed, quantum mechanics is generally unable to assign values for properties prior to measurement at all.

## Multiverse

In Dublin in 1952 Erwin Schrödinger gave a section in which at one point he jocularly warned his audience that what he was about to say might "seem lunatic". It was that, when his Nobel equations seem to be describing several different histories, they are "not alternatives but all really happen simultaneously". This is the earliest known reference to the multiverse.

## Interpretation of the Wave Function

The Schrödinger equation provides a way to calculate the wave function of a system and how it changes dynamically in time. However, the Schrödinger equation does not directly say *what*, exactly, the wave function is. Interpretations of quantum mechanics address questions such as what the relation is between the wave function, the underlying reality, and the results of experimental measurements.

An important aspect is the relationship between the Schrödinger equation and wavefunction collapse. In the oldest Copenhagen interpretation, particles follow the Schrödinger equation *except* during wavefunction collapse, during which they behave entirely differently. The advent of quantum decoherence theory allowed alternative approaches (such as the Everett many-worlds interpretation and consistent histories), wherein the Schrödinger equation is *always* satisfied, and wavefunction collapse should be explained as a consequence of the Schrödinger equation.

## Historical Background and Development

Erwin Schrödinger

Following Max Planck's quantization of light Albert Einstein interpreted Planck's quanta to be photons, particles of light, and proposed that the energy of a photon is proportional to its frequency, one of the first signs of wave–particle duality. Since energy and momentum are related in the same way as frequency and wavenumber in special relativity, it followed that the momentum $p$ of a photon is inversely proportional to its wavelength $\lambda$, or proportional to its wavenumber $k$.

$$p = \frac{h}{\lambda} = \hbar k$$

where $h$ is Planck's constant. Louis de Broglie hypothesized that this is true for all particles, even particles which have mass such as electrons. He showed that, assuming that the matter waves propagate along with their particle counterparts, electrons form standing waves, meaning that only certain discrete rotational frequencies about the nucleus of an atom are allowed. These quantized orbits correspond to discrete energy levels, and de Broglie reproduced the Bohr model formula for the energy levels. The Bohr model was based on the assumed quantization of angular momentum $L$ according to:

$$L = n\frac{h}{2\pi} = n\hbar.$$

According to de Broglie the electron is described by a wave and a whole number of wavelengths must fit along the circumference of the electron's orbit:

$$n\lambda = 2\pi r.$$

This approach essentially confined the electron wave in one dimension, along a circular orbit of radius $r$.

In 1921, prior to de Broglie, Arthur C. Lunn at the University of Chicago had used the same argument based on the completion of the relativistic energy–momentum 4-vector to derive what we now call the de Broglie relation. Unlike de Broglie, Lunn went on to formulate the differential equation now known as the Schrödinger equation, and solve for its energy eigenvalues for the hydrogen atom. Unfortunately the paper was rejected by the Physical Review, as recounted by Kamen.

Following up on de Broglie's ideas, physicist Peter Debye made an offhand comment that if particles behaved as waves, they should satisfy some sort of wave equation. Inspired by Debye's remark, Schrödinger decided to find a proper 3-dimensional wave equation for the electron. He was guided by William R. Hamilton's analogy between mechanics and optics, encoded in the observation that the zero-wavelength limit of optics resembles a mechanical system—the trajectories of light rays become sharp tracks that obey Fermat's principle, an analog of the principle of least action. A modern version of his reasoning is reproduced below. The equation he found is:

$$i\hbar\frac{\partial}{\partial t}\Psi(\mathbf{r},t) = -\frac{\hbar^2}{2m}\nabla^2\Psi(\mathbf{r},t) + V(\mathbf{r})\Psi(\mathbf{r},t).$$

However, by that time, Arnold Sommerfeld had refined the Bohr model with relativistic corrections. Schrödinger used the relativistic energy momentum relation to find what is now known as the Klein–Gordon equation in a Coulomb potential (in natural units):

$$\left(E+\frac{e^2}{r}\right)^2 \psi(x) = -\nabla^2\psi(x) + m^2\psi(x).$$

He found the standing waves of this relativistic equation, but the relativistic corrections disagreed with Sommerfeld's formula. Discouraged, he put away his calculations and secluded himself in an isolated mountain cabin in December 1925.

While at the cabin, Schrödinger decided that his earlier nonrelativistic calculations were novel enough to publish, and decided to leave off the problem of relativistic corrections for the future. Despite the difficulties in solving the differential equation for hydrogen (he had sought help from his friend the mathematician Hermann Weyl) Schrödinger showed that his nonrelativistic version of the wave equation produced the correct spectral energies of hydrogen in a paper published in 1926. In the equation, Schrödinger computed the hydrogen spectral series by treating a hydrogen atom's electron as a wave $\Psi(x, t)$, moving in a potential well $V$, created by the proton. This computation accurately reproduced the energy levels of the Bohr model. In a paper, Schrödinger himself explained this equation as follows:

This 1926 paper was enthusiastically endorsed by Einstein, who saw the matter-waves as an intuitive depiction of nature, as opposed to Heisenberg's matrix mechanics, which he considered overly formal.

The Schrödinger equation details the behavior of $\Psi$ but says nothing of its *nature*. Schrödinger tried to interpret it as a charge density in his fourth paper, but he was unsuccessful. In 1926, just a few days after Schrödinger's fourth and final paper was published, Max Born successfully interpreted $\Psi$ as the probability amplitude, whose absolute square is equal to probability density. Schrödinger, though, always opposed a statistical or probabilistic approach, with its associated discontinuities—much like Einstein, who believed that quantum mechanics was a statistical approximation to an underlying deterministic theory—and never reconciled with the Copenhagen interpretation.

Louis de Broglie in his later years proposed a real valued wave function connected to the complex wave function by a proportionality constant and developed the De Broglie–Bohm theory.

## The Wave Equation for Particles

The Schrödinger equation is a diffusion equation, the *solutions* are functions which describe wave-like motions. Wave equations in physics can normally be derived from other physical laws – the wave equation for mechanical vibrations on strings and in matter can be derived from Newton's laws – where the wave function represents the displacement of matter, and electromagnetic waves from Maxwell's equations, where the wave functions are electric and magnetic fields. The basis for Schrödinger's equation, on the other hand, is the energy of the system and a separate postulate of quantum mechanics: the wave function is a description of the system. The Schrödinger equation is therefore a new concept in itself; as Feynman put it:

The foundation of the equation is structured to be a linear differential equation based on classical energy conservation, and consistent with the De Broglie relations. The solution is the wave function $\psi$, which contains all the information that can be known about the system. In the Copenhagen interpretation, the modulus of $\psi$ is related to the probability the particles are in some spatial configuration at some instant of time. Solving the equation for $\psi$ can be used to predict how the particles will behave under the influence of the specified potential and with each other.

The Schrödinger equation was developed principally from the De Broglie hypothesis, a wave equation that would describe particles, and can be constructed as shown informally in the following sections. For a more rigorous description of Schrödinger's equation.

## Consistency with Energy Conservation

The total energy $E$ of a particle is the sum of kinetic energy $T$ and potential energy $V$, this sum is also the frequent expression for the Hamiltonian $H$ in classical mechanics:

$$E = T + V = H$$

Explicitly, for a particle in one dimension with position $x$, mass $m$ and momentum $p$, and potential energy $V$ which generally varies with position and time $t$:

$$E = \frac{p^2}{2m} + V(x,t) = H.$$

For three dimensions, the position vector **r** and momentum vector p must be used:

$$E = \frac{\mathbf{p} \cdot \mathbf{p}}{2m} + V(\mathbf{r},t) = H$$

This formalism can be extended to any fixed number of particles: the total energy of the system is then the total kinetic energies of the particles, plus the total potential energy, again the Hamiltonian. However, there can be interactions between the particles (an $N$-body problem), so the potential energy $V$ can change as the spatial configuration of particles changes, and possibly with time. The potential energy, in general, is *not* the sum of the separate potential energies for each particle, it is a function of all the spatial positions of the particles. Explicitly:

$$E = \sum_{n=1}^{N} \frac{\mathbf{p}_n \cdot \mathbf{p}_n}{2m_n} + V(\mathbf{r}_1, \mathbf{r}_2 \cdots \mathbf{r}_N, t) = H$$

## Linearity

The simplest wavefunction is a plane wave of the form:

$$\Psi(\mathbf{r},t) = Ae^{i(\mathbf{k}\cdot\mathbf{r} - \omega t)}$$

where the $A$ is the amplitude, k the wavevector, and $\omega$ the angular frequency, of the plane wave. In general, physical situations are not purely described by plane waves, so for generality the superposition principle is required; any wave can be made by superposition of sinusoidal plane waves. So if the equation is linear, a linear combination of plane waves is also an allowed solution. Hence a necessary and separate requirement is that the Schrödinger equation is a linear differential equation.

For discrete k the sum is a superposition of plane waves:

$$\Psi(\mathbf{r},t)=\sum_{n=1}^{\infty} A_n e^{i(\mathbf{k}_n\cdot\mathbf{r}-\omega_n t)}$$

for some real amplitude coefficients $A_n$, and for continuous k the sum becomes an integral, the Fourier transform of a momentum space wavefunction:

$$\Psi(\mathbf{r},t)=\frac{1}{(\sqrt{2\pi})^3}\int \Phi(\mathbf{k})e^{i(\mathbf{k}\cdot\mathbf{r}-\omega t)}d^3\mathbf{k}$$

where $d^3k = dk_x dk_y dk_z$ is the differential volume element in k-space, and the integrals are taken over all k-space. The momentum wavefunction $\Phi(k)$ arises in the integrand since the position and momentum space wavefunctions are Fourier transforms of each other.

## Consistency with the De Broglie relations

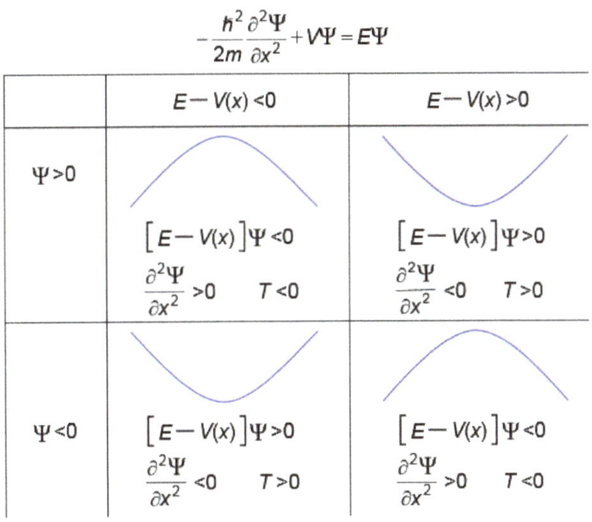


Diagrammatic summary of the quantities related to the wavefunction, as used in De broglie's hypothesis and development of the Schrödinger equation

Einstein's light quanta hypothesis (1905) states that the energy $E$ of a photon is proportional to the frequency $\nu$ (or angular frequency, $\omega = 2\pi\nu$) of the corresponding quantum wavepacket of light:

$$E = h\nu = \hbar\omega$$

Likewise De Broglie's hypothesis (1924) states that any particle can be associated with a wave, and that the momentum $p$ of the particle is inversely proportional to the wavelength $\lambda$ of such a wave (or proportional to the wavenumber, $k = 2\pi/\lambda$), in one dimension, by:

$$p=\frac{h}{\lambda}=\hbar k\ ,$$

while in three dimensions, wavelength $\lambda$ is related to the magnitude of the wavevector $\mathbf{k}$:

$$\mathbf{p} = \hbar\mathbf{k}, \quad |\mathbf{k}| = \frac{2\pi}{\lambda}.$$

The Planck–Einstein and de Broglie relations illuminate the deep connections between energy with time, and space with momentum, and express wave–particle duality. In practice, natural units comprising $\hbar = 1$ are used, as the De Broglie *equations* reduce to *identities*: allowing momentum, wavenumber, energy and frequency to be used interchangeably, to prevent duplication of quantities, and reduce the number of dimensions of related quantities.

Schrödinger's insight, late in 1925, was to express the phase of a plane wave as a complex phase factor using these relations:

$$\Psi = Ae^{i(\mathbf{k}\cdot\mathbf{r}-\omega t)} = Ae^{i(\mathbf{p}\cdot\mathbf{r}-Et)/\hbar}$$

and to realize that the first order partial derivatives were:

with respect to space:

$$\nabla\Psi = \frac{i}{\hbar}\mathbf{p}Ae^{i(\mathbf{p}\cdot\mathbf{r}-Et)/\hbar} = \frac{i}{\hbar}\mathbf{p}\Psi$$

with respect to time:

$$\frac{\partial\Psi}{\partial t} = -\frac{iE}{\hbar}Ae^{i(\mathbf{p}\cdot\mathbf{r}-Et)/\hbar} = -\frac{iE}{\hbar}\Psi$$

Another postulate of quantum mechanics is that all observables are represented by linear Hermitian operators which act on the wavefunction, and the eigenvalues of the operator are the values the observable takes. The previous derivatives are consistent with the energy operator, corresponding to the time derivative,

$$\hat{E}\Psi = i\hbar\frac{\partial}{\partial t}\Psi = E\Psi$$

where $E$ are the energy eigenvalues, and the momentum operator, corresponding to the spatial derivatives (the gradient $\nabla$),

$$\hat{\mathbf{p}}\Psi = -i\hbar\nabla\Psi = \mathbf{p}\Psi$$

where p is a vector of the momentum eigenvalues. In the above, the "hats" ( ^ ) indicate these observables are operators, not simply ordinary numbers or vectors. The energy and momentum operators are *differential operators*, while the potential energy function $V$ is just a multiplicative factor.

Substituting the energy and momentum operators into the classical energy conservation equation obtains the operator:

$$E = \frac{\mathbf{p}\cdot\mathbf{p}}{2m} + V \quad \rightarrow \quad \hat{E} = \frac{\hat{\mathbf{p}}\cdot\hat{\mathbf{p}}}{2m} + V$$

so in terms of derivatives with respect to time and space, acting this operator on the wavefunction $\Psi$ immediately led Schrödinger to his equation:

$$i\hbar\frac{\partial\Psi}{\partial t} = -\frac{\hbar^2}{2m}\nabla^2\Psi + V\Psi$$

Wave–particle duality can be assessed from these equations as follows. The kinetic energy $T$ is related to the square of momentum p. As the particle's momentum increases, the kinetic energy increases more rapidly, but since the wavenumber $|\mathrm{k}|$ increases the wavelength $\lambda$ decreases. In terms of ordinary scalar and vector quantities (not operators):

$$\mathbf{p}\cdot\mathbf{p} \propto \mathbf{k}\cdot\mathbf{k} \propto T \propto \frac{1}{\lambda^2}$$

The kinetic energy is also proportional to the second spatial derivatives, so it is also proportional to the magnitude of the *curvature* of the wave, in terms of operators:

$$\hat{T}\Psi = \frac{-\hbar^2}{2m}\nabla\cdot\nabla\Psi \propto \nabla^2\Psi.$$

As the curvature increases, the amplitude of the wave alternates between positive and negative more rapidly, and also shortens the wavelength. So the inverse relation between momentum and wavelength is consistent with the energy the particle has, and so the energy of the particle has a connection to a wave, all in the same mathematical formulation.

## Wave and Particle Motion

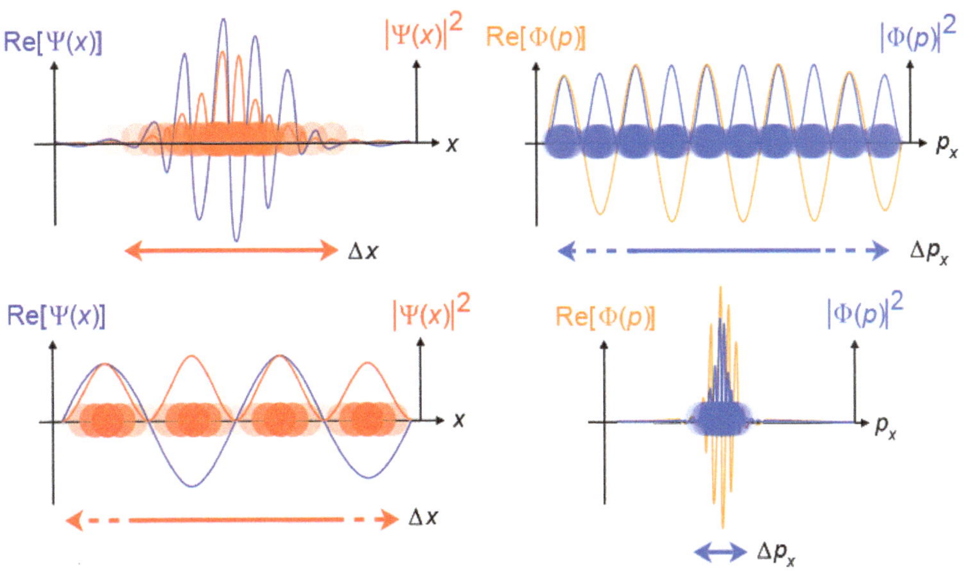


Increasing levels of wavepacket localization, meaning the particle has a more localized position

Perfect localization

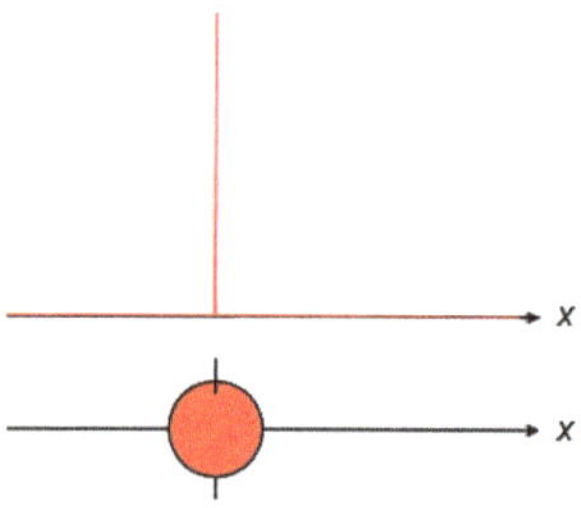


In the limit $\hbar \rightarrow 0$, the particle's position and momentum become known exactly. This is equivalent to the classical particle

Schrödinger required that a wave packet solution near position r with wavevector near k will move along the trajectory determined by classical mechanics for times short enough for the spread in k (and hence in velocity) not to substantially increase the spread in r. Since, for a given spread in k, the spread in velocity is proportional to Planck's constant ħ, it is sometimes said that in the limit as ħ approaches zero, the equations of classical mechanics are restored from quantum mechanics. Great care is required in how that limit is taken, and in what cases.

The limiting short-wavelength is equivalent to $\hbar$ tending to zero because this is limiting case of increasing the wave packet localization to the definite position of the particle. Using the Heisenberg uncertainty principle for position and momentum, the products of uncertainty in position and momentum become zero as $\hbar \rightarrow 0$:

$$\sigma(x)\sigma(p_x) \geqslant \frac{\hbar}{2} \quad \rightarrow \quad \sigma(x)\sigma(p_x) \geqslant 0$$

where $\sigma$ denotes the (root mean square) measurement uncertainty in $x$ and $p_x$ (and similarly for the $y$ and $z$ directions) which implies the position and momentum can only be known to arbitrary precision in this limit.

The Schrödinger equation in its general form

$$i\hbar \frac{\partial}{\partial t} \Psi(\mathbf{r},t) = \hat{H}\Psi(\mathbf{r},t)$$

is closely related to the Hamilton–Jacobi equation (HJE)

$$\frac{\partial}{\partial t} S(q_i,t) = H\left(q_i, \frac{\partial S}{\partial q_i}, t\right)$$

where $S$ is action and $H$ is the Hamiltonian function (not operator). Here the generalized coordinates $q_i$ for $i = 1, 2, 3$ (used in the context of the HJE) can be set to the position in Cartesian coordinates as $\mathbf{r} = (q_1, q_2, q_3) = (x, y, z)$.

Substituting

$$\Psi = \sqrt{\rho(\mathbf{r},t)}e^{iS(\mathbf{r},t)/\hbar}$$

where $\rho$ is the probability density, into the Schrödinger equation and then taking the limit $\hbar \to 0$ in the resulting equation, yields the Hamilton–Jacobi equation.

The implications are:

- The motion of a particle, described by a (short-wavelength) wave packet solution to the Schrödinger equation, is also described by the Hamilton–Jacobi equation of motion.
- The Schrödinger equation includes the wavefunction, so its wave packet solution implies the position of a (quantum) particle is fuzzily spread out in wave fronts. On the contrary, the Hamilton–Jacobi equation applies to a (classical) particle of definite position and momentum, instead the position and momentum at all times (the trajectory) are deterministic and can be simultaneously known.

## Nonrelativistic Quantum Mechanics

The quantum mechanics of particles without accounting for the effects of special relativity, for example particles propagating at speeds much less than light, is known as nonrelativistic quantum mechanics. Following are several forms of Schrödinger's equation in this context for different situations: time independence and dependence, one and three spatial dimensions, and one and $N$ particles.

In actuality, the particles constituting the system do not have the numerical labels used in theory. The language of mathematics forces us to label the positions of particles one way or another, otherwise there would be confusion between symbols representing which variables are for which particle.

### Time Independent

If the Hamiltonian is not an explicit function of time, the equation is separable into a product of spatial and temporal parts. In general, the wavefunction takes the form:

$$\Psi(\text{space coords}, t) = \psi(\text{space coords})\tau(t).$$

where $\psi$(space coords) is a function of all the spatial coordinate(s) of the particle(s) constituting the system only, and $\tau(t)$ is a function of time only.

Substituting for $\psi$ into the Schrödinger equation for the relevant number of particles in the relevant number of dimensions, solving by separation of variables implies the general solution of the time-dependent equation has the form:

$$\Psi(\text{space coords}, t) = \psi(\text{space coords})e^{-iEt/\hbar}.$$

Since the time dependent phase factor is always the same, only the spatial part needs to be solved for in time independent problems. Additionally, the energy operator $\hat{E} = i\hbar\frac{\partial}{\partial t}$ can always be replaced by the energy eigenvalue $E$, thus the time independent Schrödinger equation is an eigenvalue equation for the Hamiltonian operator:

$$\hat{H}\psi = E\psi$$

This is true for any number of particles in any number of dimensions (in a time independent potential). This case describes the standing wave solutions of the time-dependent equation, which are the states with definite energy (instead of a probability distribution of different energies). In physics, these standing waves are called "stationary states" or "energy eigenstates"; in chemistry they are called "atomic orbitals" or "molecular orbitals". Superpositions of energy eigenstates change their properties according to the relative phases between the energy levels.

The energy eigenvalues from this equation form a discrete spectrum of values, so mathematically energy must be quantized. More specifically, the energy eigenstates form a basis – any wavefunction may be written as a sum over the discrete energy states or an integral over continuous energy states, or more generally as an integral over a measure. This is the spectral theorem in mathematics, and in a finite state space it is just a statement of the completeness of the eigenvectors of a Hermitian matrix.

## One-dimensional Examples

For a particle in one dimension, the Hamiltonian is:

$$\hat{H} = \frac{\hat{p}^2}{2m} + V(x), \quad \hat{p} = -i\hbar\frac{d}{dx}$$

and substituting this into the general Schrödinger equation gives:

$$-\frac{\hbar^2}{2m}\frac{d^2}{dx^2}\psi(x) + V(x)\psi(x) = E\psi(x)$$

This is the only case the Schrödinger equation is an ordinary differential equation, rather than a partial differential equation. The general solutions are always of the form:

$$\Psi(x,t) = \psi(x)e^{-iEt/\hbar}.$$

For $N$ particles in one dimension, the Hamiltonian is:

$$\hat{H} = \sum_{n=1}^{N}\frac{\hat{p}_n^2}{2m_n} + V(x_1, x_2, \cdots x_N), \quad \hat{p}_n = -i\hbar\frac{\partial}{\partial x_n}$$

where the position of particle $n$ is $x_n$. The corresponding Schrödinger equation is:

$$-\frac{\hbar^2}{2}\sum_{n=1}^{N}\frac{1}{m_n}\frac{\partial^2}{\partial x_n^2}\psi(x_1, x_2, \cdots x_N) + V(x_1, x_2, \cdots x_N)\psi(x_1, x_2, \cdots x_N) = E\psi(x_1, x_2, \cdots x_N).$$

so the general solutions have the form:

$$\Psi(x_1, x_2, \cdots x_N, t) = e^{-iEt/\hbar}\psi(x_1, x_2 \cdots x_N)$$

For non-interacting distinguishable particles, the potential of the system only influences each particle separately, so the total potential energy is the sum of potential energies for each particle:

$$V(x_1, x_2, \cdots x_N) = \sum_{n=1}^{N} V(x_n).$$

and the wavefunction can be written as a product of the wavefunctions for each particle:

$$\Psi(x_1, x_2, \cdots x_N, t) = e^{-iEt/\hbar} \prod_{n=1}^{N} \psi(x_n),$$

For non-interacting identical particles, the potential is still a sum, but wavefunction is a bit more complicated – it is a sum over the permutations of products of the separate wavefunctions to account for particle exchange. In general for interacting particles, the above decompositions are *not* possible.

## Free Particle

For no potential, $V = 0$, so the particle is free and the equation reads:

$$-E\psi = \frac{\hbar^2}{2m}\frac{d^2\psi}{dx^2}$$

which has oscillatory solutions for $E > 0$ (the $C_n$ are arbitrary constants):

$$\psi_E(x) = C_1 e^{i\sqrt{2mE/\hbar^2}\,x} + C_2 e^{-i\sqrt{2mE/\hbar^2}\,x}$$

and exponential solutions for $E < 0$

$$\psi_{-|E|}(x) = C_1 e^{\sqrt{2m|E|/\hbar^2}\,x} + C_2 e^{-\sqrt{2m|E|/\hbar^2}\,x}.$$

The exponentially growing solutions have an infinite norm, and are not physical. They are not allowed in a finite volume with periodic or fixed boundary conditions.

## Constant Potential

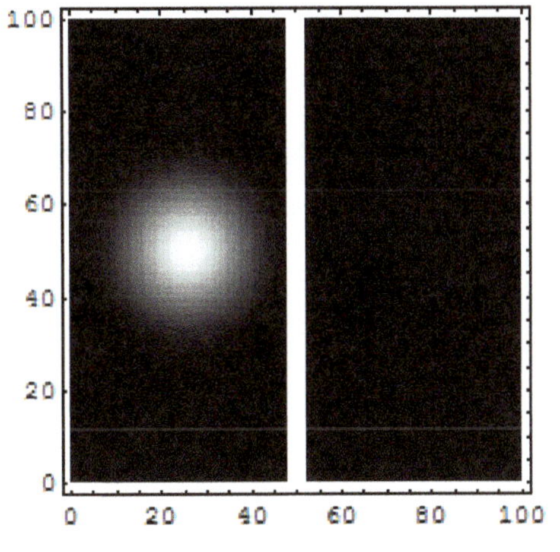


For a constant potential, $V = V_0$, the solution is oscillatory for $E > V_0$ and exponential for $E < V_0$, corresponding to energies that are allowed or disallowed in classical mechanics. Oscillatory solutions have a classically allowed energy and correspond to actual classical motions, while the exponential solutions have a disallowed energy and describe a small amount of quantum bleeding into the classically disallowed region, due to quantum tunneling. If the potential $V_0$ grows to infinity, the motion is classically confined to a finite region. Viewed far enough away, every solution is reduced to an exponential; the condition that the exponential is decreasing restricts the energy levels to a discrete set, called the allowed energies.

## Harmonic Oscillator

The Schrödinger equation for this situation is

$$E\psi = -\frac{\hbar^2}{2m}\frac{d^2}{dx^2}\psi + \frac{1}{2}m\omega^2 x^2 \psi$$

It is a notable quantum system to solve for; since the solutions are exact (but complicated – in terms of Hermite polynomials), and it can describe or at least approximate a wide variety of other systems, including vibrating atoms, molecules, and atoms or ions in lattices, and approximating other potentials near equilibrium points. It is also the basis of perturbation methods in quantum mechanics.

There is a family of solutions – in the position basis they are

$$\psi_n(x) = \sqrt{\frac{1}{2^n n!}} \cdot \left(\frac{m\omega}{\pi\hbar}\right)^{1/4} \cdot e^{-\frac{m\omega x^2}{2\hbar}} \cdot H_n\left(\sqrt{\frac{m\omega}{\hbar}}x\right)$$

where $n = 0,1,2,...$, and the functions $H_n$ are the Hermite polynomials.

## Three-dimensional Examples

The extension from one dimension to three dimensions is straightforward, all position and momentum operators are replaced by their three-dimensional expressions and the partial derivative with respect to space is replaced by the gradient operator.

The Hamiltonian for one particle in three dimensions is:

$$\hat{H} = \frac{\hat{\mathbf{p}} \cdot \hat{\mathbf{p}}}{2m} + V(\mathbf{r}), \quad \hat{\mathbf{p}} = -i\hbar\nabla$$

generating the equation:

$$-\frac{\hbar^2}{2m}\nabla^2\psi(\mathbf{r}) + V(\mathbf{r})\psi(\mathbf{r}) = E\psi(\mathbf{r})$$

with stationary state solutions of the form:

$$\Psi(\mathbf{r},t) = \psi(\mathbf{r})e^{-iEt/\hbar}$$

where the position of the particle is **r**. Two useful coordinate systems for solving the Schrödinger equation are Cartesian coordinates so that $\mathbf{r} = (x, y, z)$ and spherical polar coordinates so that $\mathbf{r} = (r, \theta, \varphi)$, although other orthogonal coordinates are useful for solving the equation for systems with certain geometric symmetries.

For $N$ particles in three dimensions, the Hamiltonian is:

$$\hat{H} = \sum_{n=1}^{N} \frac{\hat{\mathbf{p}}_n \cdot \hat{\mathbf{p}}_n}{2m_n} + V(\mathbf{r}_1, \mathbf{r}_2, \cdots \mathbf{r}_N), \quad \hat{\mathbf{p}}_n = -i\hbar\nabla_n$$

where the position of particle $n$ is $\mathbf{r}_n$ and the gradient operators are partial derivatives with respect to the particle's position coordinates. In Cartesian coordinates, for particle $n$, the position vector is $\mathbf{r}_n = (x_n, y_n, z_n)$ while the gradient and Laplacian operator are respectively:

$$\nabla_n = \mathbf{e}_x \frac{\partial}{\partial x_n} + \mathbf{e}_y \frac{\partial}{\partial y_n} + \mathbf{e}_z \frac{\partial}{\partial z_n}, \quad \nabla_n^2 = \nabla_n \cdot \nabla_n = \frac{\partial^2}{\partial {x_n}^2} + \frac{\partial^2}{\partial {y_n}^2} + \frac{\partial^2}{\partial {z_n}^2}$$

The Schrödinger equation is:

$$-\frac{\hbar^2}{2} \sum_{n=1}^{N} \frac{1}{m_n} \nabla_n^2 \Psi(\mathbf{r}_1, \mathbf{r}_2, \cdots \mathbf{r}_N) + V(\mathbf{r}_1, \mathbf{r}_2, \cdots \mathbf{r}_N) \Psi(\mathbf{r}_1, \mathbf{r}_2, \cdots \mathbf{r}_N) = E\Psi(\mathbf{r}_1, \mathbf{r}_2, \cdots \mathbf{r}_N)$$

with stationary state solutions:

$$\Psi(\mathbf{r}_1, \mathbf{r}_2 \cdots \mathbf{r}_N, t) = e^{-iEt/\hbar} \psi(\mathbf{r}_1, \mathbf{r}_2 \cdots \mathbf{r}_N)$$

Again, for non-interacting distinguishable particles the potential is the sum of particle potentials

$$V(\mathbf{r}_1, \mathbf{r}_2, \cdots \mathbf{r}_N) = \sum_{n=1}^{N} V(\mathbf{r}_n)$$

and the wavefunction is a product of the particle wavefuntions

$$\Psi(\mathbf{r}_1, \mathbf{r}_2 \cdots \mathbf{r}_N, t) = e^{-iEt/\hbar} \prod_{n=1}^{N} \psi(\mathbf{r}_n).$$

For non-interacting identical particles, the potential is a sum but the wavefunction is a sum over permutations of products. The previous two equations do not apply to interacting particles.

Following are examples where exact solutions are known.

## Hydrogen Atom

This form of the Schrödinger equation can be applied to the hydrogen atom:

$$E\psi = -\frac{\hbar^2}{2\mu} \nabla^2 \psi - \frac{e^2}{4\pi\varepsilon_0 r} \psi$$

where $e$ is the electron charge, r is the position of the electron ($r = |\mathrm{r}|$ is the magnitude of the position), the potential term is due to the Coulomb interaction, wherein $\varepsilon_o$ is the electric constant (permittivity of free space) and

$$\mu = \frac{m_e m_p}{m_e + m_p}$$

is the 2-body reduced mass of the hydrogen nucleus (just a proton) of mass $m_p$ and the electron of mass $m_e$. The negative sign arises in the potential term since the proton and electron are oppositely charged. The reduced mass in place of the electron mass is used since the electron and proton together orbit each other about a common centre of mass, and constitute a two-body problem to solve. The motion of the electron is of principle interest here, so the equivalent one-body problem is the motion of the electron using the reduced mass.

The wavefunction for hydrogen is a function of the electron's coordinates, and in fact can be separated into functions of each coordinate. Usually this is done in spherical polar coordinates:

$$\psi(r,\theta,\phi) = R(r)Y_\ell^m(\theta,\phi) = R(r)\Theta(\theta)\Phi(\phi)$$

where $R$ are radial functions and $Y_\ell^m(\theta,\varphi)$ are spherical harmonics of degree $\ell$ and order $m$. This is the only atom for which the Schrödinger equation has been solved for exactly. Multi-electron atoms require approximative methods. The family of solutions are:

$$\psi_{n\ell m}(r,\theta,\phi) = \sqrt{\left(\frac{2}{na_0}\right)^3 \frac{(n-\ell-1)!}{2n[(n+\ell)!]}} e^{-r/na_0} \left(\frac{2r}{na_0}\right)^\ell L_{n-\ell-1}^{2\ell+1}\left(\frac{2r}{na_0}\right) \cdot Y_\ell^m(\theta,\phi)$$

where:

- $a_0 = \dfrac{4\pi\varepsilon_0 \hbar^2}{m_e e^2}$ is the Bohr radius,
- $L_{n-\ell-1}^{2\ell+1}(\cdots)$ are the generalized Laguerre polynomials of degree $n - \ell - 1$.
- $n$, $\ell$, $m$ are the principal, azimuthal, and magnetic quantum numbers respectively: which take the values:

  $n = 1,2,3,\ldots \quad \ell = 0,1,2,\ldots,n-1 \quad m = -\ell,\ldots,\ell$

NB: generalized Laguerre polynomials are defined differently by different authors.

## Two-electron Atoms or Ions

The equation for any two-electron system, such as the neutral helium atom (He, $Z = 2$), the negative hydrogen ion ($H^-$, $Z = 1$), or the positive lithium ion ($Li^+$, $Z = 3$) is:

$$E\psi = -\hbar^2\left[\frac{1}{2\mu}\left(\nabla_1^2 + \nabla_2^2\right) + \frac{1}{M}\nabla_1 \cdot \nabla_2\right]\psi + \frac{e^2}{4\pi\varepsilon_0}\left[\frac{1}{r_{12}} - Z\left(\frac{1}{r_1} + \frac{1}{r_2}\right)\right]\psi$$

where $\mathrm{r}_1$ is the position of one electron ($r_1 = |\mathrm{r}_1|$ is its magnitude), $\mathrm{r}_2$ is the position of the other electron ($r_2 = |\mathrm{r}_2|$ is the magnitude), $r_{12} = |\mathrm{r}_{12}|$ is the magnitude of the separation between them given by

$$|\mathbf{r}_{12}| = |\mathbf{r}_2 - \mathbf{r}_1|$$

$\mu$ is again the two-body reduced mass of an electron with respect to the nucleus of mass $M$, so this time

$$\mu = \frac{m_e M}{m_e + M}$$

and $Z$ is the atomic number for the element (not a quantum number).

The cross-term of two laplacians

$$\frac{1}{M}\nabla_1 \cdot \nabla_2$$

is known as the *mass polarization term*, which arises due to the motion of atomic nuclei. The wavefunction is a function of the two electron's positions:

$$\psi = \psi(\mathbf{r}_1, \mathbf{r}_2).$$

There is no closed form solution for this equation.

## Time Dependent

This is the equation of motion for the quantum state. In the most general form, it is written:

$$i\hbar \frac{\partial}{\partial t}\Psi = \hat{H}\Psi.$$

and the solution, the wavefunction, is a function of all the particle coordinates of the system and time. Following are specific cases.

For one particle in one dimension, the Hamiltonian

$$\hat{H} = \frac{\hat{p}^2}{2m} + V(x,t), \quad \hat{p} = -i\hbar \frac{\partial}{\partial x}$$

generates the equation:

$$i\hbar \frac{\partial}{\partial t}\Psi(x,t) = -\frac{\hbar^2}{2m}\frac{\partial^2}{\partial x^2}\Psi(x,t) + V(x,t)\Psi(x,t)$$

For $N$ particles in one dimension, the Hamiltonian is:

$$\hat{H} = \sum_{n=1}^{N} \frac{\hat{p}_n^2}{2m_n} + V(x_1, x_2, \cdots x_N, t), \quad \hat{p}_n = -i\hbar \frac{\partial}{\partial x_n}$$

where the position of particle $n$ is $x_n$, generating the equation:

$$i\hbar\frac{\partial}{\partial t}\Psi(x_1,x_2\cdots x_N,t)=-\frac{\hbar^2}{2}\sum_{n=1}^{N}\frac{1}{m_n}\frac{\partial^2}{\partial x_n^2}\Psi(x_1,x_2\cdots x_N,t)+V(x_1,x_2\cdots x_N,t)\Psi(x_1,x_2\cdots x_N,t).$$

For one particle in three dimensions, the Hamiltonian is:

$$\hat{H}=\frac{\hat{\mathbf{p}}\cdot\hat{\mathbf{p}}}{2m}+V(\mathbf{r},t),\quad \hat{\mathbf{p}}=-i\hbar\nabla$$

generating the equation:

$$i\hbar\frac{\partial}{\partial t}\Psi(\mathbf{r},t)=-\frac{\hbar^2}{2m}\nabla^2\Psi(\mathbf{r},t)+V(\mathbf{r},t)\Psi(\mathbf{r},t)$$

For $N$ particles in three dimensions, the Hamiltonian is:

$$\hat{H}=\sum_{n=1}^{N}\frac{\hat{\mathbf{p}}_n\cdot\hat{\mathbf{p}}_n}{2m_n}+V(\mathbf{r}_1,\mathbf{r}_2,\cdots\mathbf{r}_N,t),\quad \hat{\mathbf{p}}_n=-i\hbar\nabla_n$$

where the position of particle $n$ is $\mathbf{r}_n$, generating the equation:

$$i\hbar\frac{\partial}{\partial t}\Psi(\mathbf{r}_1,\mathbf{r}_2,\cdots\mathbf{r}_N,t)=-\frac{\hbar^2}{2}\sum_{n=1}^{N}\frac{1}{m_n}\nabla_n^2\Psi(\mathbf{r}_1,\mathbf{r}_2,\cdots\mathbf{r}_N,t)+V(\mathbf{r}_1,\mathbf{r}_2,\cdots\mathbf{r}_N,t)\Psi(\mathbf{r}_1,\mathbf{r}_2,\cdots\mathbf{r}_N,t)$$

This last equation is in a very high dimension, so the solutions are not easy to visualize.

## Solution Methods

| General techniques: | Methods for special cases: |
|---|---|
| • Perturbation theory<br>• The variational method<br>• Quantum Monte Carlo methods<br>• Density functional theory<br>• The WKB approximation and semi-classical expansion | • List of quantum-mechanical systems with analytical solutions<br>• Hartree–Fock method and post Hartree–Fock methods |

## Properties

The Schrödinger equation has the following properties: some are useful, but there are shortcomings. Ultimately, these properties arise from the Hamiltonian used, and the solutions to the equation.

## Linearity

In the development above, the Schrödinger equation was made to be linear for generality, though this has other implications. If two wave functions $\psi_1$ and $\psi_2$ are solutions, then so is any linear combination of the two:

$$\psi = a\psi_1 + b\psi_2$$

where $a$ and $b$ are any complex numbers (the sum can be extended for any number of wavefunctions). This property allows superpositions of quantum states to be solutions of the Schrödinger equation. Even more generally, it holds that a general solution to the Schrödinger equation can be found by taking a weighted sum over all single state solutions achievable. For example, consider a wave function $\Psi(x, t)$ such that the wave function is a product of two functions: one time independent, and one time dependent. If states of definite energy found using the time independent Schrödinger equation are given by $\psi_E(x)$ with amplitude $A_n$ and time dependent phase factor is given by

$$e^{-iE_n t/\hbar},$$

then a valid general solution is

$$\Psi(x,t) = \sum_n A_n \psi_{E_n}(x) e^{-iE_n t/\hbar}.$$

Additionally, the ability to scale solutions allows one to solve for a wave function without normalizing it first. If one has a set of normalized solutions $\psi_n$, then

$$\Psi = \sum_n A_n \psi_n$$

can be normalized by ensuring that

$$\sum_n |A_n|^2 = 1.$$

This is much more convenient than having to verify that

$$\int_{-\infty}^{\infty} |\Psi(x)|^2 dx = \int_{-\infty}^{\infty} \Psi(x)\Psi^*(x) dx = 1.$$

## Real Energy Eigenstates

For the time-independent equation, an additional feature of linearity follows: if two wave functions $\psi_1$ and $\psi_2$ are solutions to the time-independent equation with the same energy $E$, then so is any linear combination:

$$\hat{H}(a\psi_1 + b\psi_2) = a\hat{H}\psi_1 + b\hat{H}\psi_2 = E(a\psi_1 + b\psi_2).$$

Two different solutions with the same energy are called *degenerate*.

In an arbitrary potential, if a wave function $\psi$ solves the time-independent equation, so does its complex conjugate, denoted $\psi^*$. By taking linear combinations, the real and imaginary parts of $\psi$ are each solutions. If there is no degeneracy they can only differ by a factor.

In the time-dependent equation, complex conjugate waves move in opposite directions. If $\Psi(x, t)$ is one solution, then so is $\Psi^*(x, -t)$. The symmetry of complex conjugation is called time-reversal symmetry.

## Space and Time Derivatives

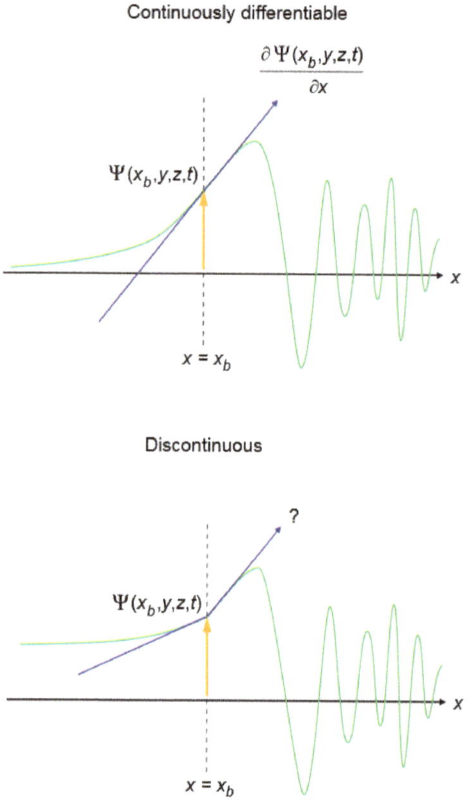


Continuity of the wavefunction and its first spatial derivative (in the $x$ direction, $y$ and $z$ coordinates not shown), at some time $t$

The Schrödinger equation is first order in time and second in space, which describes the time evolution of a quantum state (meaning it determines the future amplitude from the present).

Explicitly for one particle in 3-dimensional Cartesian coordinates – the equation is

$$i\hbar\frac{\partial\Psi}{\partial t}=-\frac{\hbar^2}{2m}\left(\frac{\partial^2\Psi}{\partial x^2}+\frac{\partial^2\Psi}{\partial y^2}+\frac{\partial^2\Psi}{\partial z^2}\right)+V(x,y,z,t)\Psi.$$

The first time partial derivative implies the initial value (at $t = 0$) of the wavefunction

$$\Psi(x,y,z,0)$$

is an arbitrary constant. Likewise – the second order derivatives with respect to space implies the wavefunction *and* its first order spatial derivatives.

$$\Psi(x_b, y_b, z_b, t)$$

$$\frac{\partial}{\partial x}\Psi(x_b, y_b, z_b, t) \quad \frac{\partial}{\partial y}\Psi(x_b, y_b, z_b, t) \quad \frac{\partial}{\partial z}\Psi(x_b, y_b, z_b, t)$$

are all arbitrary constants at a given set of points, where $x_b$, $y_b$, $z_b$ are a set of points describing boundary $b$ (derivatives are evaluated at the boundaries). Typically there are one or two boundaries, such as the step potential and particle in a box respectively.

As the first order derivatives are arbitrary, the wavefunction can be a continuously differentiable function of space, since at any boundary the gradient of the wavefunction can be matched.

On the contrary, wave equations in physics are usually *second order in time*, notable are the family of classical wave equations and the quantum Klein–Gordon equation.

## Local Conservation of Probability

The Schrödinger equation is consistent with probability conservation. Multiplying the Schrödinger equation on the right by the complex conjugate wavefunction, and multiplying the wavefunction to the left of the complex conjugate of the Schrödinger equation, and subtracting, gives the continuity equation for probability:

$$\frac{\partial}{\partial t}\rho(\mathbf{r},t)+\nabla\cdot\mathbf{j}=0,$$

where

$$\rho=|\Psi|^2=\Psi^*(\mathbf{r},t)\Psi(\mathbf{r},t)$$

is the probability density (probability per unit volume, * denotes complex conjugate), and

$$\mathbf{j}=\frac{1}{2m}\left(\Psi^*\hat{\mathbf{p}}\Psi-\Psi\hat{\mathbf{p}}\Psi^*\right)$$

is the probability current (flow per unit area).

Hence predictions from the Schrödinger equation do not violate probability conservation.

## Positive Energy

If the potential is bounded from below, meaning there is a minimum value of potential energy, the eigenfunctions of the Schrödinger equation have energy which is also bounded from below. This can be seen most easily by using the variational principle, as follows.

For any linear operator $\hat{A}$ bounded from below, the eigenvector with the smallest eigenvalue is the vector $\psi$ that minimizes the quantity

$$\langle\psi|\hat{A}|\psi\rangle$$

over all $\psi$ which are normalized. In this way, the smallest eigenvalue is expressed through the variational principle. For the Schrödinger Hamiltonian $\hat{H}$ bounded from below, the smallest eigenvalue is called the ground state energy. That energy is the minimum value of

$$\langle\psi\,|\,\hat{H}\,|\,\psi\rangle = \int \psi^*(\mathbf{r})\left[-\frac{\hbar^2}{2m}\nabla^2\psi(\mathbf{r}) + V(\mathbf{r})\psi(\mathbf{r})\right]d^3\mathbf{r} = \int\left[\frac{\hbar^2}{2m}|\nabla\psi|^2 + V(\mathbf{r})\,|\psi|^2\right]d^3\mathbf{r} = \langle\hat{H}\rangle$$

(using integration by parts). Due to the complex modulus of $\psi^2$ (which is positive definite), the right hand side always greater than the lowest value of $V(x)$. In particular, the ground state energy is positive when $V(x)$ is everywhere positive.

For potentials which are bounded below and are not infinite over a region, there is a ground state which minimizes the integral above. This lowest energy wavefunction is real and positive definite – meaning the wavefunction can increase and decrease, but is positive for all positions. It physically cannot be negative: if it were, smoothing out the bends at the sign change (to minimize the wavefunction) rapidly reduces the gradient contribution to the integral and hence the kinetic energy, while the potential energy changes linearly and less quickly. The kinetic and potential energy are both changing at different rates, so the total energy is not constant, which can't happen (conservation). The solutions are consistent with Schrödinger equation if this wavefunction is positive definite.

The lack of sign changes also shows that the ground state is nondegenerate, since if there were two ground states with common energy $E$, not proportional to each other, there would be a linear combination of the two that would also be a ground state resulting in a zero solution.

## Analytic Continuation To Diffusion

The above properties (positive definiteness of energy) allow the analytic continuation of the Schrödinger equation to be identified as a stochastic process. This can be interpreted as the Huygens–Fresnel principle applied to De Broglie waves; the spreading wavefronts are diffusive probability amplitudes. For a free particle (not subject to a potential) in a random walk, substituting $\tau = it$ into the time-dependent Schrödinger equation gives:

$$\frac{\partial}{\partial\tau}X(\mathbf{r},\tau) = \frac{\hbar}{2m}\nabla^2 X(\mathbf{r},\tau), \quad X(\mathbf{r},\tau) = \Psi(\mathbf{r},\tau/i)$$

which has the same form as the diffusion equation, with diffusion coefficient $\hbar/2m$. In that case, the diffusivity yields the De Broglie relation in accordance with the Markov process.

## Regularity

On the space $L^2$ of square-integrable densities, the Schrödinger semigroup $e^{it\hat{H}}$ is a unitary evolution, and therefore surjective. The flows satisfy the Schrödinger equation $i\partial_t u = \widehat{H}u$, where the derivative is taken in the distribution sense. However, since $\widehat{H}$ for most physically reasonable Hamiltonians (e.g., the Laplace operator, possibly modified by a potential) is unbounded in $L^2$, this shows that the semigroup flows lack Sobolev regularity in general. Instead, solutions of the Schrödinger equation satisfy a Strichartz estimate.

## Relativistic Quantum Mechanics

Relativistic quantum mechanics is obtained where quantum mechanics and special relativity simultaneously apply. In general, one wishes to build relativistic wave equations from the relativistic energy–momentum relation

$$E^2 = (pc)^2 + (m_0c^2)^2,$$

instead of classical energy equations. The Klein–Gordon equation and the Dirac equation are two such equations. The Klein–Gordon equation,

$$\frac{1}{c^2}\frac{\partial^2}{\partial t^2}\psi - \nabla^2\psi + \frac{m^2c^2}{\hbar^2}\psi = 0.$$

,

was the first such equation to be obtained, even before the nonrelativistic one, and applies to massive spinless particles. The Dirac equation arose from taking the "square root" of the Klein–Gordon equation by factorizing the entire relativistic wave operator into a product of two operators – one of these is the operator for the entire Dirac equation.

The general form of the Schrödinger equation remains true in relativity, but the Hamiltonian is less obvious. For example, the Dirac Hamiltonian for a particle of mass $m$ and electric charge $q$ in an electromagnetic field (described by the electromagnetic potentials $\varphi$ and A) is:

$$\hat{H}_{\text{Dirac}} = \gamma^0\left[c\gamma\cdot\left(\hat{\mathbf{p}} - q\mathbf{A}\right) + mc^2 + \gamma^0 q\phi\right],$$

in which the $\gamma = (\gamma^1, \gamma^2, \gamma^3)$ and $\gamma^0$ are the Dirac gamma matrices related to the spin of the particle. The Dirac equation is true for all spin-$1/2$ particles, and the solutions to the equation are 4-component spinor fields with two components corresponding to the particle and the other two for the antiparticle.

For the Klein–Gordon equation, the general form of the Schrödinger equation is inconvenient to use, and in practice the Hamiltonian is not expressed in an analogous way to the Dirac Hamiltonian. The equations for relativistic quantum fields can be obtained in other ways, such as starting from a Lagrangian density and using the Euler–Lagrange equations for fields, or use the representation theory of the Lorentz group in which certain representations can be used to fix the equation for a free particle of given spin (and mass).

In general, the Hamiltonian to be substituted in the general Schrödinger equation is not just a function of the position and momentum operators (and possibly time), but also of spin matrices. Also, the solutions to a relativistic wave equation, for a massive particle of spin $s$, are complex-valued $2(2s + 1)$-component spinor fields.

## Quantum Field Theory

The general equation is also valid and used in quantum field theory, both in relativistic and nonrelativistic situations. However, the solution $\psi$ is no longer interpreted as a "wave", but should be interpreted as an operator acting on states existing in a Fock space.

## First Order Form

The Schrödinger equation can also be derived from a first order form similar to the manner in which the Klein-Gordon equation can be derived from the Dirac equation. In 1D the first order equation is given by

$$-i\partial_z \psi = (i\eta \partial_t + \eta^\dagger m)\psi$$

This equation allows for the inclusion of spin in nonrelativistic quantum mechanics. Squaring the above equation yields the Schrödinger equation in 1D. The matrices $\eta$ obey the following properties

$$\eta^2 = 0 \quad (\eta^\dagger)^2 = 0 \quad \{\eta, \eta^\dagger\} = 2I$$

The 3 dimensional version of the equation is given by

$$-i\gamma_i \partial_i \psi = (i\eta \partial_t + \eta^\dagger m)\psi$$

Here $\eta = (\gamma_0 + i\gamma_5)/\sqrt{2}$ is a $4\times4$ nilpotent matrix and $\gamma_i$ are the Dirac gamma matrices ($i = 1, 2, 3$). The Schrödinger equation in 3D can be obtained by squaring the above equation. In the nonrelativistic limit $E - m \simeq E'$ and $E + m \simeq 2m$, the above equation can be derived from the Dirac equation.

## Particle in an Infinite Potential Well

Particle of mass m and fixed total energy E confined to a relatively small segment of one dimensional space between x = 0 and x = a. In terms of the potential energy we can view that the particle is trapped in an infinitely deep one dimensional potential well with $U = 0$ constant for 0 < x < a

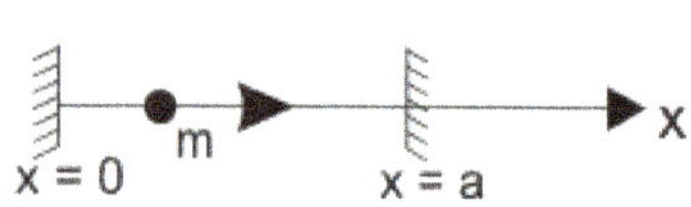


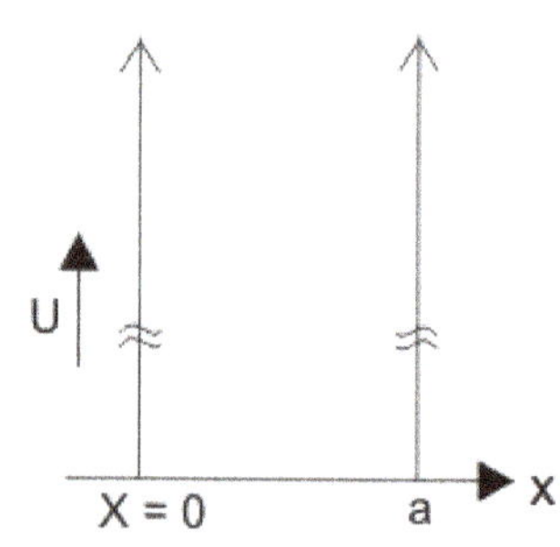


Now time independent Schrödinger equation is

$$\frac{d^2\psi}{dx^2} + k^2\psi = 0 \qquad 0 < x < a$$

$$k = \sqrt{\left(\frac{2mE}{\hbar^2}\right)} \quad or \quad E = \frac{\hbar^2 k^2}{2m}$$

Boundary conditions

$$\psi(0) = 0 \; as \quad \psi = 0 \; for \; x < 0$$
$$\psi(a) = 0 \; as \quad \psi = 0 \; for \; x > a$$

Solving $\psi(x) = A\sin kx + B\cos kx$ which can also be represented as

$$A + \exp(+jkx) + A - \exp(-jkx)$$
$$\psi(0) = 0 \Rightarrow B = 0.$$

$$\psi(a) = 0 \Rightarrow A\sin ka = 0.$$

Since $A \neq 0$, as $\psi$ vanishes for all x, so

$\rightarrow ka = n\pi, n = \pm1, \pm2, \pm3$ ........discrete values.

$$\rightarrow k = n\pi / a = k_n$$

Therefore, $\psi(x) = \psi_n(x) = A_n \sin\left(\frac{n\pi x}{a}\right) \rightarrow$ modes or eigen functions

for $k_n = \frac{n\pi}{a}$, Energy $E = E_n = \frac{\hbar^2 n^2 \pi^2}{2ma^2}$

So Particle can have only a Few Discrete Energy States/Levels

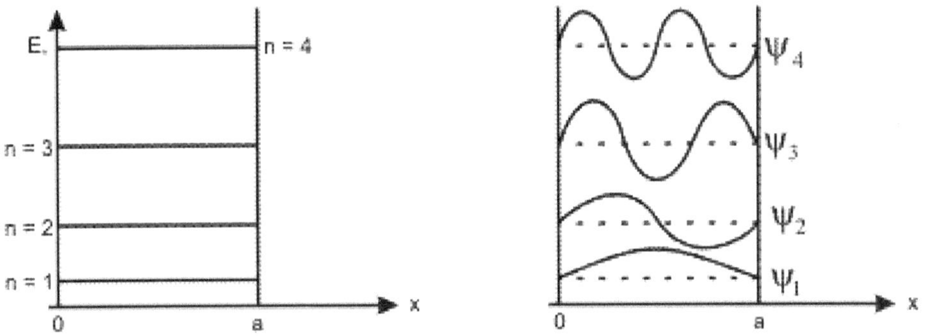


Wave function $\psi_n(x) = A_n \sin\left(\frac{n\pi x}{a}\right)$

$\rightarrow$ standing wave

$\rightarrow$ particle is bouncing back and forth between the walls of the potential well.

$\rightarrow < px >$ average value of momentum at any x is o. So to calculate momentum we have to isolate forward wave going in +x or backward wave going in -x.

For +x wave, $p_{x_+} = \frac{n\pi\hbar}{a}$ and -x going wave $p_{x_-} = \frac{-n\pi\hbar}{a}$ Discrete $E_{-p_x}$ points lie on the continuous $E = \frac{\langle p_x \rangle^2}{2m}$ parabola of a free particle.

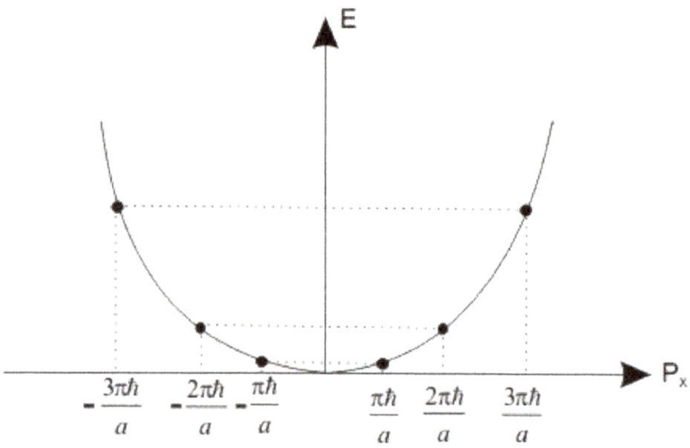


The integer n is called the quantum number

Value of $A_n$ from the equation that $\int_v \psi_n^* \psi_n dv = 1 \rightarrow A_n = \sqrt{\frac{2}{a}}$

## Finite Potential Well

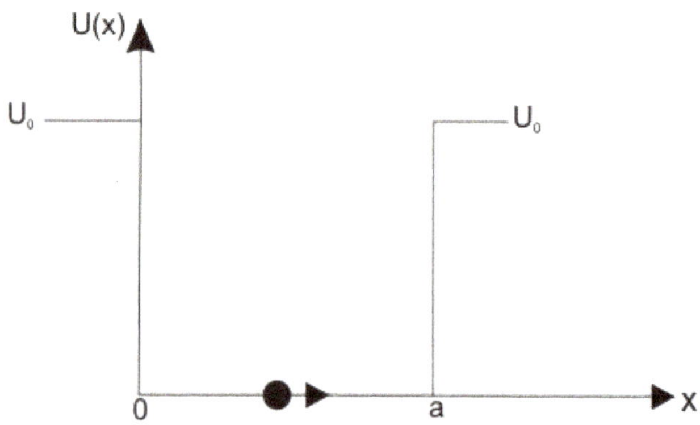


U(x) = 0 for 0 < x < a

U(x) = U 0 otherwise

We assume a particle with energy $E < U_0$ confined to the potential well

For 0 < x < a,

$$\frac{d^2\psi_0}{dx^2} + k^2\psi_0 = 0, k = \sqrt{\frac{2mE}{\hbar^2}}$$

For x <0,

$$\frac{d^2\psi_-}{dx^2} - \alpha^2\psi_- = 0, \alpha = \sqrt{\frac{2m}{\hbar^2}(U_o - E)}$$

For x > 0,

$$\frac{d^2\psi_+}{dx^2} - \alpha^2\psi_+ = 0$$

The three sets of solution are

$$\psi_-(x) = A_- e^{\alpha x} + B_- e^{-\alpha x}, \qquad x < 0$$
$$\psi_0(x) = A_0 \sin kx + B_0 \cos kx, \qquad 0 < x < a$$
$$\psi_+(x) = A_+ e^{\alpha x} + B_+ e^{-\alpha x}, \qquad x > a$$

Now we set up the boundary conditions.

Far from the potential well, wave functions must go to zero as the probability of finding the particle away from potential region = 0

Therefore, $\psi_-(-\infty) = 0, \psi_+(+\infty) = 0$

$\psi$ must be continuous at well boundary

$$\psi_-(0) = \psi_0(0), \psi_+(a) = \psi_0(a)$$

$\frac{d\psi}{dx}$ must be continuous at well boundaries x = 0 & x = a

$$\left.\frac{d\psi_-}{dx}\right|_0 = \left.\frac{d\psi}{dx}\right|_0 \;\&\; \left.\frac{d\psi_+}{dx}\right|_a = \left.\frac{d\psi_0}{dx}\right|_a$$

From boundary conditions $B_- = 0, A_+ = 0$

And we get four simultaneous equations

$$A_- = B_0$$
$$A_0 \sin ka + B_0 \cos ka = B_+ e^{-\alpha a}$$
$$aA_- = kA_0$$
$$kA_0 \cos ka - kB_0 \sin ka = -\alpha B_+ e^{-\alpha a}$$

elimination $A_-, B_0$ and $B_+$ we obtain

$$A_0 \left\lfloor (k^2 - \alpha^2)\sin ka - 2\alpha k \cos ka \right\rfloor = 0$$

if $A_0 = 0, \psi = 0$ identically.

Therefore, $(k^2 - \alpha^2)\sin ka - 2\alpha k \cos ka = 0$

Or

$$\tan ka = \frac{2\alpha k}{k^2 - \alpha^2}$$

This is a transcendental equation

We define a dimensionless quantity

$$f_0 = \sqrt{\frac{2mU_0}{h^2}}$$

$f_0$ is constant of the system.

And

$$\xi = \frac{E}{U_0} \qquad (0 < \xi < 1)$$

$\alpha = f_0\sqrt{1-\xi}, k = f_0\sqrt{\xi}$ from definition of k & $\alpha$.

Eigen value equation becomes

$$\tan(f_0 a\sqrt{\xi}) = \frac{2\sqrt{\xi(1-\xi)}}{2\xi - 1} = S(\xi)$$

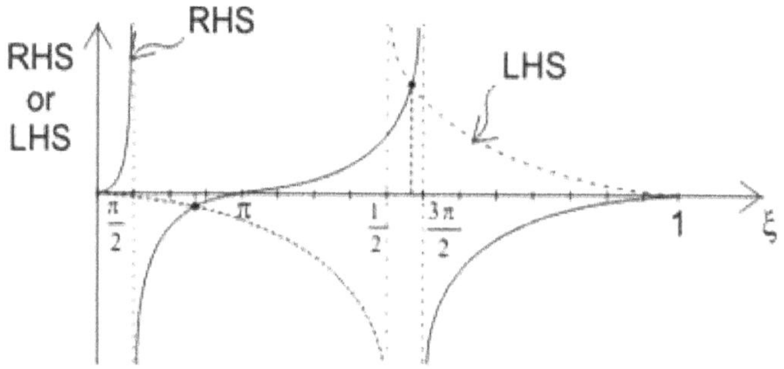


Let $f_0 a\sqrt{\xi} = \theta$ then

if $f_0 a < \pi, \tan\theta$ intersect $S(\xi)$ at one points

$\pi < f_0 a < 2\pi, \tan\theta S(\xi)$ intersect at 2 points

$2\pi < f_0 a < 3\pi, \tan\theta$ intersect $S(\xi)$ at 3 points

So for,

$m\pi < f_0 a < (m+a)\pi$ there are m points.

In the limit $U_0 \to \infty$ (E finite). $S(\xi) = 0$.

$\tan ka = 0$

$$\to k = k_n = \frac{n\pi}{a}, \qquad n = \pm 1, \pm 2 \ldots$$

Finite potential

In finite potential wells, we talked about Eigen value equation

$$\tan(f_0 a\sqrt{\xi}) = \frac{2\sqrt{\xi(1-\xi)}}{2\xi - 1}$$

for a given system constant

$$f_0 a = a\sqrt{\frac{2mU_0}{\hbar^2}}$$

We have a discrete no. of solutions where LHS or RHS of above Equ.

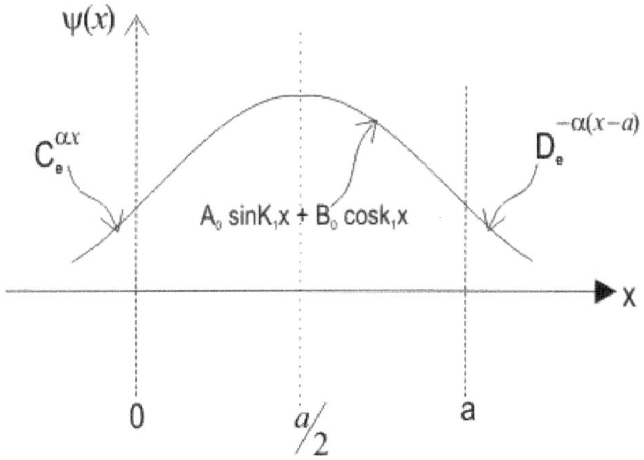


Wave Function of Finite Potential Wells

For each intersection a value of $\xi_1 = \frac{E_1}{U_0} \rightarrow$ particular $\alpha_1$ and $k_1$

$\psi \neq 0$ outside the well of potential

One can clearly see from this that there is a finite probability of existence outside in the classically forbidden region (classically a particle with an energy $E \quad U$ cannot exist outside the well). If the potential well is slightly modified as in the figure below.

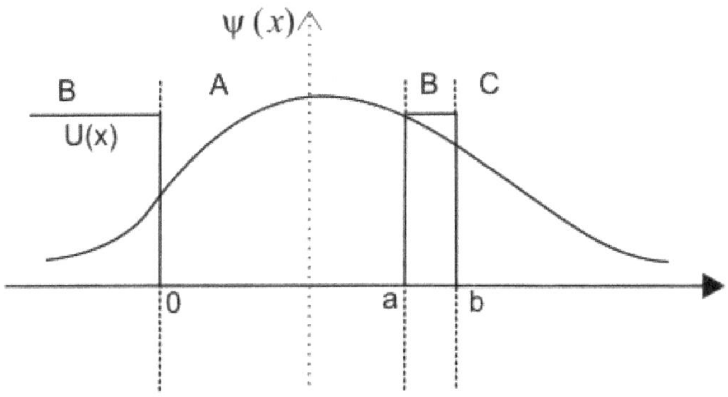


The wave function will be nonzero in region C also.

Thus the particle will have a finite probability of existing or coming in region C past the potential barrier B

In C the particle can appear as a free particle.

This quantum mechanical phenomenon of passing through a barrier is known as tunneling. The wave function is different inside and outside.

Therefore there exists a finite probability of reflection at the well walls, called quantum mechanical reflection at the well walls. In analogy to optics it may be looked at as two partially reflecting mirrors, where an infinite potential well could be visualized as two 100% reflecting mirrors.

It is used in Tunnel diodes and operation of many other solid-state devices. One of the usage of this phenomenon for high frequency decive is called Resonant Tunnelling Diode (RTD), which would be used as an oscillator and even as an amplifier.

## Linear Harmonic Oscillator

A harmonic oscillator is a particle which is bound to an equilibrium position by a force which is proportional to the displacement from that position.

Thus we have,

$$force = -\gamma x = -\frac{dU}{dx}$$

where $\gamma$ is the spring constant.

The potential is expressed as,

$$U(x) = \frac{1}{2}\gamma x^2$$

The linear harmonic oscillator can then be visualized on a mass connected to a spring of spring constant $\gamma$ on shown in Figure.

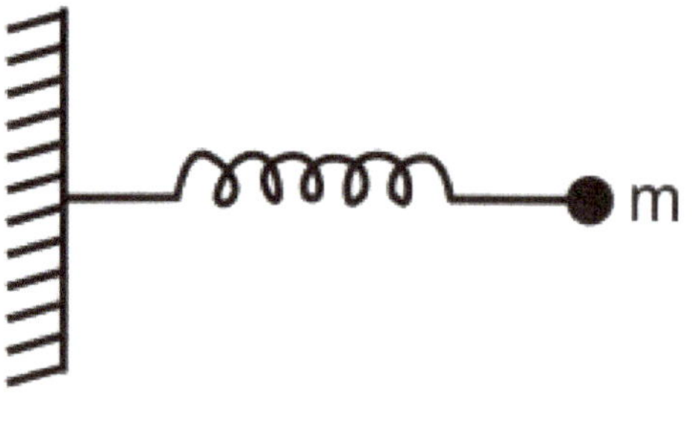


The time independent Schrödinger equation is given by,

$$\frac{\hbar^2}{2m}\frac{d^2\psi}{dx^2} + \frac{1}{2}\gamma x^2\psi = E\psi$$

or,

$$\frac{d^2\psi}{dx^2}+\frac{2m}{\hbar^2}\left(E-\frac{1}{2}\gamma x^2\right)\psi=0$$

To solve above equation, we consider a dimension less quantity,

$$y=\left(\frac{m\gamma}{\hbar^2}\right)^{1/4}x$$

Or,

$$\lambda=\frac{2}{\hbar}\left(\frac{m}{\gamma}\right)^{1/2}E$$

using above two equations,

$$\frac{d^2\psi}{dy^2}+(\lambda-y^2)\psi=0$$

For large values of y we can neglect $\lambda$ we get from above equation,

$$\frac{d^2\psi}{dy^2}-y^2\psi=0$$

Above equation is satisfied approximately by the solution,

$$\psi(y)=e^{-y^2/2}$$

Substituting above two equations we get,

$$\frac{d^2\psi}{dy^2}+(1-y^2)\psi=0$$

Hence we consider the exact solution as,

$$\psi(y)=e^{-y^2/2}\ \xi(y)$$

Putting the value of $\psi(y)$ from above equation to $\frac{d^2\psi}{dy^2}+(\lambda-y^2)\psi=0$ equation,

$$\frac{d^2\xi}{dy^2}-2y\frac{d\xi}{dy}+(\lambda-1)\xi=0$$

The trick next, is to linearize the above equation.

Above equation can be solved by using the power series method.

Let the trail solution be,

$$\xi(y)=\sum_{n=0}^{\infty}a_n y^n$$

$$\frac{d^2\xi}{dy^2}=\sum_{n=0}^{\infty}(n+1)(n+2)a_{n+}2y^n$$

$$-2y\frac{d\xi}{dy}=\sum_{n=0}^{\infty}-2na_n y^n$$

$$(\lambda-1)\xi=\sum_{n=0}^{\infty}(\lambda-1)a_n y^n$$

Putting these equations in above equation,

$$\sum_{n=0}^{\infty}[(n+1)(n+2)a_{n+2}-(\lambda-1-2n)a_n]y^n=0$$

This equation must hold for all values of $\xi$, and therefore the coefficient of each power of $\xi$ must vanish separately.

This gives in the recursion relation between $a_{n+2}$ and a n ,

$$a_{n+2}=\frac{2n-\lambda+1}{(n+1)(n+2)}a_n$$

It seems that knowing $a_0$ and $a_1 a_1, a_3, a_4, a5$.......... can be calculated by using above equation,

$$\rightarrow a_2=-\frac{(\lambda-1)}{2!}a_0 \qquad a_3=-\frac{(\lambda-3)}{3!}a_1$$

$$\rightarrow a_4=+\frac{(\lambda-1)(\lambda-5)}{4!}a_0 \qquad a_5=+\frac{(\lambda-3)(\lambda-7)}{5!}a_1$$

Thus we can write $\xi(y)=\sum_{n=0}^{\infty}a_n y^n$ equation as,

$$\xi(y)=a_0\left[1-\frac{(\lambda-1)}{2!}y^2+\frac{(\lambda-1)(\lambda-5)}{4!}y^4+....\right]+a_1\left[y-\frac{(\lambda-3)}{3!}+\frac{(\lambda-3)}{5!}y^5+....\right]$$

If in the equation $\frac{d^2\psi}{dy^2}+(\lambda-y^2)\psi=0$, $\lambda-1-2n$ should be zero for some value of the index n, then $a_{n+2}=0$.

But since $a_{n+4}$ is a multiple of $a_{n+2}$ so on, all the succeeding coefficients which are related to $a_n$ by the recursion relation would vanish, and one or the other bracketed series in above equation would terminate to become a polynomial of degree n.

This occurs, when,

$$\lambda - 1 - 2n = 0$$

$$or, \qquad \lambda = 2n + 1 (n = 0, 1, 2 ..)$$

## Energy Quantization

We have obtained the condition when the wave function is acceptable as

$$\lambda = 2n + 1_{(n=0,1,2...)}$$

again $\lambda$ was $\lambda = \frac{2}{\hbar}\left(\frac{m}{y}\right)^{1/2} E$

$$E_n = \frac{\hbar}{2}\omega(2n+1) \qquad \left[As \quad \omega = \left(\frac{y}{m}\right)^{1/2}\right]$$

$$E_n = \left(n + \frac{1}{2}\right)\hbar\omega$$

The variation of the energy levels is shown in the Figure.

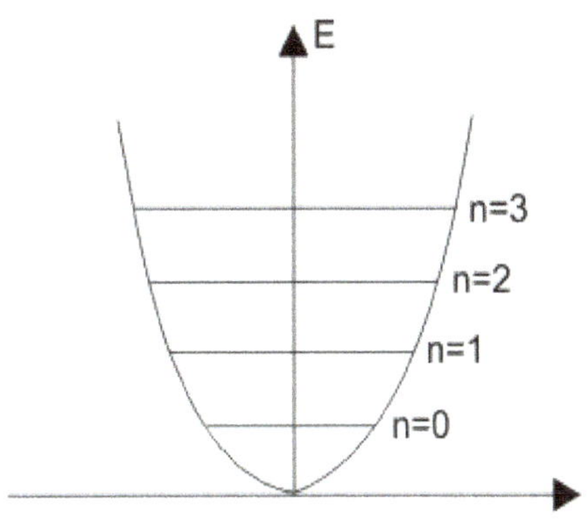

3

# Energy Bands and Energy Carriers in Semiconductors

Charge carriers are those particles found in electrical conductors that are free to move. Semi conductors require a certain correlation between valence band and the conduction band in order for a charge carrier to carry electric current. In order to completely understand energy bands and energy carriers it is necessary to understand the processes related to it. The following chapter elucidates the varied processes and mechanisms associated with this area of study.

## Bonding Forces and Energy Bands in Solids

- Electrons are restricted to sets of discrete energy levels within atoms, with large gaps among them where no energy state is available for the electron to occupy.
- Electrons in solids also are restricted to certain energies and are not allowed at other energies.
- Difference $\rightarrow$ in the solid, the electron has a range (or band) of available energies.
- The discrete energy levels of the isolated atom spread into bands of energies in the solid because
  i) in the solid, the wave functions of electrons in neighboring atoms overlap, thus, it affects the potential energy term and the boundary conditions in the schrondinger equation, and different energies are obtained in the solution, and
  ii) an electron is not necessarily localized at a particular atom.
- The influence of neighboring atoms on the energy levels of a particular atom can be treated as a small perturbation, giving rise to shifting and splitting of energy states into energy bands.

### Bonding Forces in Solids

### Ionic Bonding

- Example: NaCl.
- Na (Z = 11) gives up its outermost shell electron to Cl (Z=17) atom, thus the crystal is made up of ions with the electronic structures of the inert atoms Ne and Ar.
- Note: the ions have net electric charges after the electron exchange $\rightarrow Na^{+}$ ion has a net

positive charge, having lost an electron, and $Cl^-$ ion has a net negative charge, having acquired an electron.

- Thus, an electrostatic attractive force is established, and the balance is reached when this equals the net repulsive force.
- Note: all the electrons are tightly bound to the atom.
- Since there are no loosely bound electrons to participate in current flow → NaCl is a good insulator.

## Metallic Bonding

- In metals, the outer shell is filled by no more than three electrons (loosely bound and given up easily) => great chemical activity and high electrical conductivity.
- Outer electron(s) contributed to the crystal as a whole => solid made up of ions with closed shells immersed in a sea of free electrons, which are free to move about the crystal under the influence of an electric field.
- Coulomb attraction force between the ions and the electrons hold the lattice together.

## Covalent Bonding

- Exhibited by the diamond lattice semiconductors.
- Each atom surrounded by four nearest neighbors, each having four electrons in the outermost orbit.
- Each atom shares its valence electrons with its four nearest neighbors.
- Bonding forces arise from a quantum mechanical interaction between the shared electrons.
- Both electrons belong to each bond, are indistinguishable, and have opposite spins.
- No free electrons available at 0 K, however, by thermal or optical excitation, electrons can be excited out of a covalent bond and can participate in current conduction => important feature of semiconductors.

## Mixed Bonding

- Shown by III-V compounds bonding partly ionic and partly covalent.
- Ionic character of bonding becomes more prominent as the constituent atoms move further away in the periodic table, e.g., II-VI compounds.

## Energy Bands

- As isolated atoms are brought together to form a solid, the electron wave functions begin to overlap.

- Various interactions occur, and, at the proper interatomic spacing for the crystal, the forces of attraction and repulsion find a balance.
- Due to Pauli exclusion principle, the discrete energy levels of individual atoms split into bands belonging to the pair instead of to individual atoms.
- In a solid, due to large number of atoms, the split energy levels for essentially continuous bands of energy.

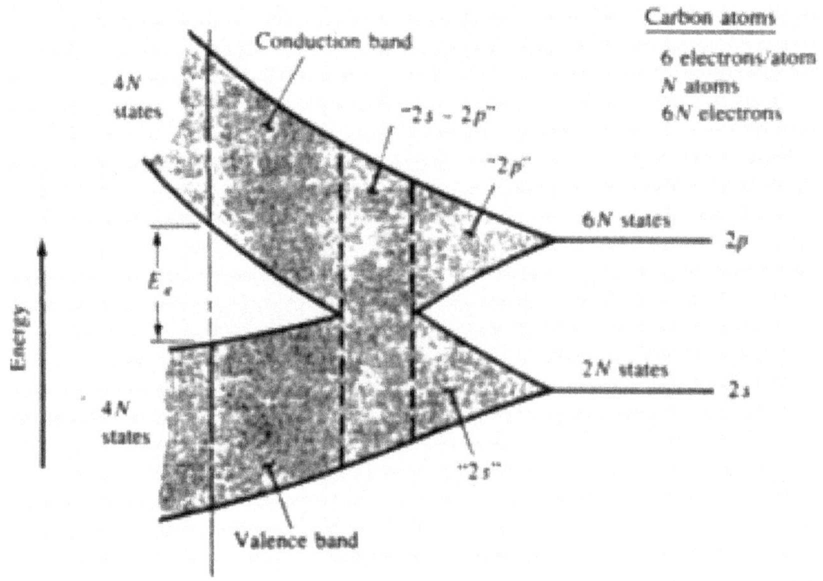


Splitting of individual energy levels to energy bands as atoms are brought closer together.

- Imaginary formation of a diamond crystal from isolated carbon atoms ($1s^2 2s^2 2p^2$).
- Each atom has two 1s states, two 2s states, six 2p states, and higher states.
- For N atoms, the numbers of states are 2N, 2N, and 6N of type 1s, 2s, and 2p respectively.
- With a reduction in the interatomic spacing, these energy levels split into bands, and the 2s and 2p bands merge into a single band having 8N available states.
- As the interatomic spacing approaches the equilibrium spacing of diamond crystal, this band splits into two bands separated by an energy gap $E_g$, where no allowed energy states for electrons exist => *forbidden gap*.
- The upper band (called the *conduction band*) and the lower band (called the *valence band*) contain 4N states each.
- The lower 1s band is filled with 2N electrons, however, the 4N electrons residing in the original n = 2 state will now occupy states either in the *valence band* or in the *conduction band.*
- At 0 K, the electrons will occupy the lowest energy states available to them => thus, the 4N states in the valence band will be completely filled, and the 4N states in the conduction band will be completely empty.

## Metals, Semiconductors and Insulators

- For electrons to move under an applied electric field, there must be states available to them.
- A completely filled band cannot contribute to current transport; neither can a completely empty band.
- Thus, semiconductors at 0 K are perfect insulators.
- With thermal or optical excitation, some of these electrons can be excited from the valence band to the conduction band, and then they can contribute to the current transport process.
- At temperatures other than 0 K, the magnitude of the band gap separates an insulator from a semiconductor, e.g., at 300 K, $E_g$ (diamond) = 5 eV (insulator), and $E_g$ (Silicon) = 1.12 eV (semiconductor).
- Number of electrons available for conduction can be increased greatly in semiconductors by reasonable amount of thermal or optical energy.
- In metals, the bands are either partially filled or they overlap => thus, electrons and empty states coexist => great electrical conductivity.

## Direct and Indirect Semiconductors

- In a typical quantitative calculation of band structures, the wave function of a single electron traveling through a perfectly periodic lattice is assumed to be in the form of a plane wave moving in the x-direction (say) with propagation constant k, also called a *wave vector*.
- In quantum mechanics, the electron momentum can be given by $p = \hbar k$
- The space dependent wave function for the electron is $\psi_k(x) = U(K_x, x)e^{jk,x}$ where the function $U(K_x, x)$ modulates the wave function according to the periodicity of the lattice.
- Allowed values of energy, while plotted as a function of k, gives the E-k diagram.
- Since the periodicity of most lattices is different in various directions, the E-k diagram is a complex surface, which is to be visualized in three dimensions.

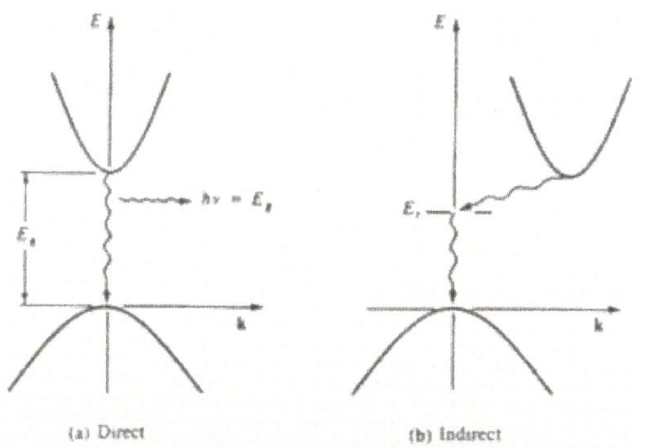


Direct and indirect transition of electrons from the conduction band to the valence band: (a) direct - with accompanying photon emission, (b) indirect via defect level.

- Direct band gap semiconductor: the minima of the conduction band and the maxima of the valence band occur at the same value of k=> an electron making the smallest energy transition from the conduction band to the valence band can do so without a change in k (and, the momentum).
- Indirect band gap semiconductor: the minima of the conduction band and the maxima of the valence band occur for different values of k, thus, the smallest energy transition for an electron requires a change in momentum.
- Electron falling from conduction band to an empty state in valence band => recombination.
- Recombination probability for direct band gap semiconductors is much higher than that for indirect band gap semiconductors.
- Direct band gap semiconductors give up the energy released during this transition (= $E_g$) in the form of light => used for optoelectronic applications (e.g., LEDs and LASERs).
- Recombination in indirect band gap semiconductors occurs through some defect states within the band gap, and the energy is released in the form of heat given to the lattice.

## Variation of Energy Bands with Alloy Composition

- The band structures of III-V ternary and quaternary compounds change as their composition is varied.
- There are three valleys in the conduction band: Γ (at k = 0), L, and X.
- In GaAs, the Γ valley has the minimum energy (direct with $E_g$ = 1.43 eV) with very few electrons residing in L and X valleys (except for high field excitations).
- In AlAs, the X valley has minimum energy (indirect with $E_g$ = 2.16 eV).

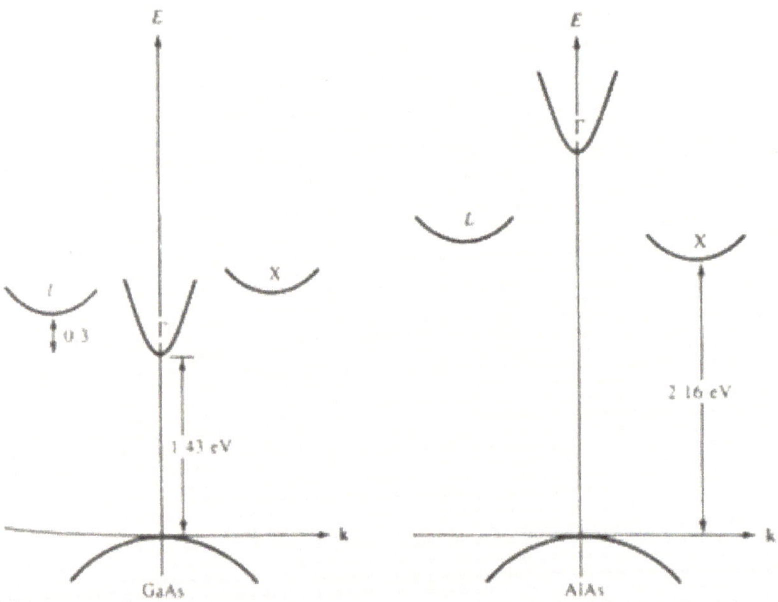


The E-k diagram of (a) GaAs and (b) AlAs, showing the three valleys (L, Γ, and X) in the conduction band.

## Charge Carriers in Semiconductors

- In a metal, the atoms are imbedded in a "sea" of free electrons, and these electrons can move as a group under the influence of an applied electric field.
- In semiconductors at 0 K, all states in the valence band are full, and all states in the conduction band are empty.
- At T > 0 K, electrons get thermally excited from the valence band to the conduction band, and contribute to the conduction process in the conduction band.
- The empty states left in the valence band can also contribute to current conduction.
- Also, introduction of impurities has an important effect on the availability of the charge carriers.
- Considerable flexibility in controlling the electrical properties of semiconductors.

## Electrons and Holes

- For T> 0 K, there would be some electrons in the otherwise empty conduction band, and some empty states in the otherwise filled valence band.
- The empty states in the valence band are referred to as holes.
- If the conduction band electron and the valence band hole are created by thermal excitation of a valence band electron to the conduction band, then they are called electron-hole pair (EHP).
- After excitation to the conduction band, an electron is surrounded by a large number of empty states, e.g., the equilibrium number of EHPs at 300 K in Si is $\sim 10^{10} / cm^3$, whereas the Si atom density is $\sim 10^{22} / cm^3$.
- Thus, the electrons in the conduction band are free to move about via the many available empty states.
- Corresponding problem of charge transport in the valence band is slightly more complex.
- Current transport in the valence band can be accounted for by keeping track of the holes themselves.
- In a filled band, all available energy states are occupied.
- For every electron moving with a given velocity, there is an equal and opposite electron motion somewhere else in the band.
- Under an applied electric field, the net current is zero, since for every electron j moving with a velocity $V_j$, there is a corresponding electron $j^t$ moving with a velocity $-V_j$.
- In a unit volume, the current density J can be given by

$$j = (-q)\sum_{i}^{N} v_i = 0 \text{ (filled band)}$$

where N is the number of $electrons / cm^3$ in the band, and q is the electronic charge.

- Now, if the $j^{th}$ electron is removed and a hole is created in the valence band, then the net current density

$$J = (-q)\sum_{i}^{N} v_i - (-q)v_j \qquad (j^{th} electron\, mis\sin g)$$
$$= (+q)v_i \qquad (\sin ce\, the\, first\, term\, is\, zero)$$

- Thus, the current contribution of the empty state (hole), obtained by removing the jth electron, is equivalent to that of a positively charged particle with velocity $v_j$ .
- Note that actually this transport is accounted for by the motion of the uncompensated electron $j^t$ having a charge of q and moving with a velocity $v_j$ .
- Its current contribution (- q)(- $v_j$ ) is equivalent to that of a positively charged particle with velocity $+v_j$ .
- For simplicity, therefore, the empty states in the valence band are called holes, and they are assigned positive charge and positive mass.
- The electron energy increases as one moves up the conduction band, and electrons gravitate downward towards the bottom of the conduction band.
- On the other hand, hole energy increases as one moves down the valence band (since holes have positive charges), and holes gravitate upwards towards the top of the valence band.

## Effective Mass

- The "wave-particle" motion of electrons in a lattice is not the same as that for a free electron, because of the interaction with the periodic potential of the lattice.
- To still be able to treat these particles as "free", the rest mass has to be altered to take into account the influence of the lattice.
- The calculation of effective mass takes into account the shape of the energy bands in three-dimensional k-space, taking appropriate averages over the various energy bands.
- The effective mass of an electron in a band with a given (E,k) relation is given by

$$m^* = \frac{[h/(2\pi)]^2}{d^2E/dk^2}$$

- Thus, the effective mass is an inverse function of the curvature of the E-k diagram: weak curvature gives large mass, and strong curvature gives small mass.
- Note that in general, the effective mass is a tensor quantity, however, for parabolic bands, it is a constant.
- Another interesting feature is that the curvature $d^2E/dk^2$ is positive at the conduction

band minima, however, it is negative at the valence band maxima.

- Thus, the electrons near the top of the valence band have negative effective mass.
- Valence band electrons with negative charge and negative mass move in an electric field in the same direction as holes with positive charge and positive mass.
- Thus, the charge transport in the valence band can be fully accounted for by considering hole motion alone.
- The electron and hole effective masses are denoted by $m_n^*$ and $m_p^*$ respectively.

The effective mass for electrons and holes.
(x $m_0$, where $m_0$ is the rest mass for electrons)

| | Ge | Si | GaAs |
|---|---|---|---|
| $m_n^*$ | 0.55 | 1.1 | 0.067 |
| $m_p^*$ | 0.37 | 0.56 | 0.48 |

## Intrinsic Material

- A perfect semiconductor crystal with no impurities or lattice defects.
- No carriers at 0 K, since the valence band is completely full and the conduction band is completely empty.
- For T > 0 K, electrons are thermally excited from the valence band to the conduction band (EHP generation).
- EHP generation takes place due to breaking of covalent bonds => required energy = $E_g$.
- The excited electron becomes free and leaves behind an empty state (hole).
- Since these carriers are created in pairs, the electron concentration ($n / cm^3$) is always equal to the hole concentration ($p / cm^3$), and each of these is commonly referred to as the *intrinsic carrier concentration* ($n_i$).
- Thus, for intrinsic material n = p = $n_i$.
- These carriers are not localized in the lattice; instead they spread out over several lattice spacings, and are given by quantum mechanical probability distributions.
- Note: ni = f(T).
- To maintain a steady-state carrier concentration, the carriers must also recombine at the same rate at which they are generated.

- Recombination occurs when an electron from the conduction band makes a transition (direct or indirect) to an empty state in the valence band, thus annihilating the pair.
- At equilibrium, $r_i = g_i$, where $g_i$ and $r_i$ are the generation and recombination rates respectively, and both of these are temperature dependent.
- $g_i$ (T) increases with temperature, and a new carrier concentration ni is established, such that the higher recombination rate $r_i$ (T) just balances generation.
- At any temperature, the rate of recombination is proportional to the equilibrium concentration of electrons and holes, and can be given by $r_i = \alpha_r n_0 p_0 = \alpha_r n_i^2 = g_i$
- where $\alpha_r$ is a constant of proportionality (depends on the mechanism by which recombination takes place).

## Extrinsic Material

- In addition to thermally generated carriers, it is possible to create carriers in the semiconductor by purposely introducing impurities into the crystal => *doping*.
- Most common technique for varying the conductivity of semiconductors.
- By doping, the crystal can be made to have predominantly electrons (n-type) or holes (p-type).
- When a crystal is doped such that the equilibrium concentrations of electrons (no) and holes (po) are different from the intrinsic carrier concentration (ni), the material is said to be *extrinsic*.
- Doping creates additional levels within the band gap.
- In Si, column V elements of the periodic table (e.g., P, As, Sb) introduce energy levels very near (typically 0.03-0.06 eV) the conduction band.
- At 0 K, these levels are filled with electrons, and very little thermal energy (50 K to 100 K) is required for these electrons to get excited to the conduction band.
- Since these levels donate electrons to the conduction band, they are referred to as the *donor* levels.
- Thus, Si doped with donor impurities can have a significant number of electrons in the conduction band even when the temperature is not sufficiently high enough for the intrinsic carriers to dominate, i.e., $n_0 >> n_i, \; p_0$ => n-type material, with electrons as *majority carriers* and holes as *minority carriers*.
- In Si, column III elements of the periodic table (e.g., B, Al, Ga, In) introduce energy levels very near (typically 0.03-0.06 eV) the valence band.
- At 0 K, these levels are empty, and very little thermal energy (50 K to 100 K) is required for electrons in the valence band to get excited to these levels, and leave behind holes in the valence band.

- Since these levels accept electrons from the valence band, they are referred to as the acceptor levels.
- Thus, Si doped with acceptor impurities can have a significant number of holes in the valence band even at a very low temperature, i.e., $p_0 >> n_i$ , $n_0$ => p-type material, with holes as majority carriers and electrons as minority carriers.
- The extra electron for column V elements is loosely bound and it can be liberated very easily=> ionization; thus, it is free to participate in current conduction.
- Similarly, column III elements create holes in the valence band, and they can also participate in current conduction.
- Rough calculation of the ionization energy can be made based on the Bohr's model for $H_2$ atoms, considering the loosely bound electron orbiting around the tightly bound core electrons. Thus,

$$E = \frac{m_n^* q^4}{2(4\pi\varepsilon_0\varepsilon_r)^2[h/(2\pi)]^2}$$ where $\varepsilon_r$ is the relative permittivity of Si.

- In III-V compounds, column VI impurities (e.g., S, Se, Te) occupying column V sites act as donors. Similarly, column II impurities (e.g., Be, Zn, Cd) occupying column III sites act as acceptors.
- When a column IV material (e.g., Si, Ge) is used to dope III-V compounds, then they may substitute column III elements (and act as donors), or substitute column V elements (and act as acceptors)$\Rightarrow$ *amphoteric dopants*.
- Doping creates a large change in the electrical conductivity, e.g., with a doping of $10^{15} / cm^3$, the resistivity of Si changes from $2\times10$ $\Omega$-cm to $5\Omega$-cm.

## Carrier Concentrations

- For the calculation of semiconductor electrical properties and analyzing device behavior, it is necessary to know the number of charge carriers/cm$^3$ in the material.
- The majority carrier concentration in a heavily doped material is obvious, since for each impurity atom, one majority carrier is obtained.
- However, the minority carrier concentration and the dependence of carrier concentrations on temperature are not obvious.
- To obtain the carrier concentrations, their distribution over the available energy states is required.
- These distributions are calculated using statistical methods.

## The Fermi Level

- Electrons in solids obey *Fermi-Dirac* (FD) statistics.

- This statistics accounts for the indistinguishability of the electrons, their wave nature, and the Pauli exclusion principle.
- The Fermi-Dirac distribution function f(E) of electrons over a range of allowed energy levels at thermal equilibrium can be given by

  $f(E) = \frac{1}{1 + e^{(E-E_F)/kT}}$ where k is Boltzmann's constant (= 8.62 x $10^{-5}$ eV/K = 1.38 x $10^{-23}$ J/K).
- This gives the probability that an available energy state at E will be occupied by an electron at an absolute temperature T.
- $E_F$ is called the Fermi level and is a measure of the average energy of the electrons in the lattice => an extremely important quantity for analysis of device behavior.
- Note: for (E - $E_F$) > 3kT (known as *Boltzmann approximation*), f(E) $\approx$ exp[ - (E- $E_F$ )/ kT] => this is referred to as the Maxwell-Boltzmann (MB) distribution (followed by gas atoms).
- The probability that an energy state at $E_F$ will be occupied by an electron is 1/2 at all temperatures.
- At 0 K, the distribution takes a simple rectangular form, with all states below $E_F$ occupied, and all states above $E_F$ empty.
- At T > 0 K, there is a finite probability of states above $E_F$ to be occupied and states below $E_F$ to be empty.
- The F-D distribution function is highly symmetric, i.e., the probability f( $E_F + \Delta E$ ) that a state E above $E_F$ is filled is the same as the probability [1- f( $E_F - \Delta E$ )] that a state E below $E_F$ is empty.
- This symmetry about EF makes the Fermi level a natural reference point for the calculation of electron and hole concentrations in the semiconductor.
- Note: f(E) is the probability of occupancy of an available state at energy E, thus, if there is no available state at E (e.g., within the band gap of a semiconductor), there is no possibility of finding an electron there.
- For intrinsic materials, the Fermi level lies close to the middle of the band gap (the difference between the effective masses of electrons and holes accounts for this small deviation from the mid gap).
- In n-type material, the electrons in the conduction band outnumber the holes in the valence band, thus, the Fermi level lies closer to the conduction band.
- Similarly, in p-type material, the holes in the valence band outnumber the electrons in the conduction band, thus, the Fermi level lies closer to the valence band.
- The probability of occupation f(E) in the conduction band and the probability of vacancy [1- f(E)] in the valence band are quite small, however, the densities of available states in

these bands are very large, thus a small change in f(E) can cause large changes in the carrier concentrations.

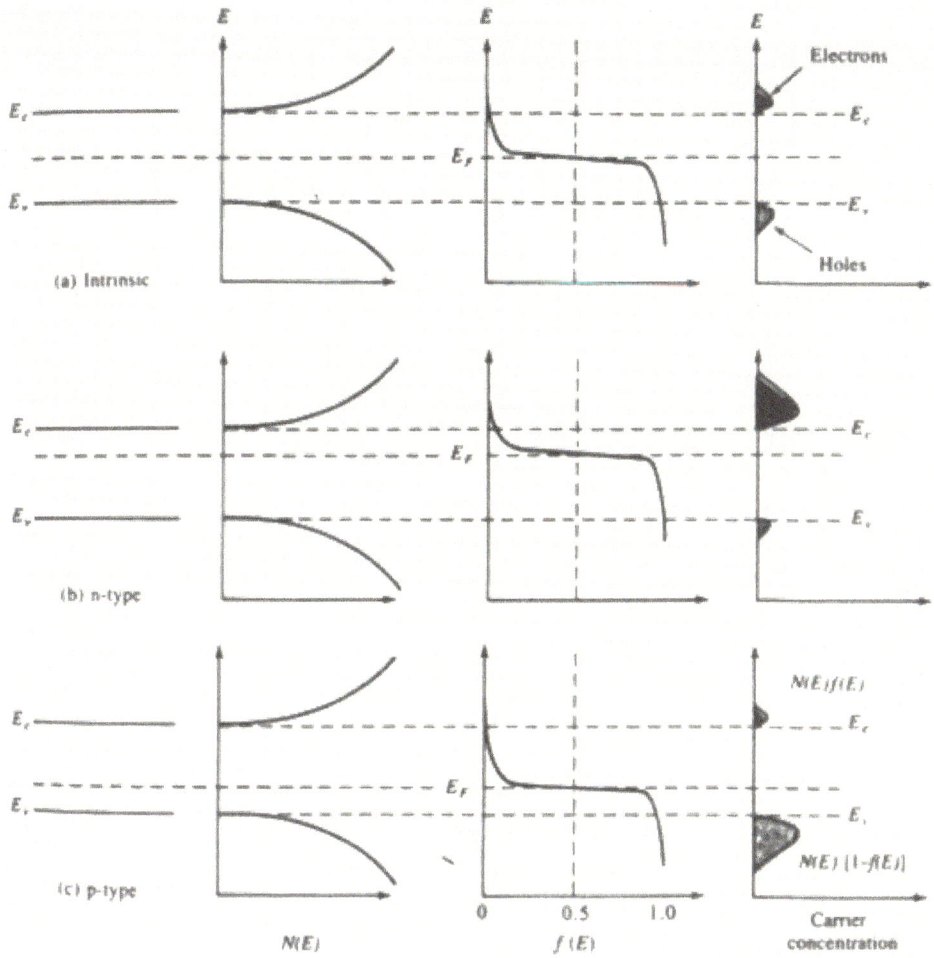


The density of states N(E), the Fermi-Dirac distribution function f(E), and the carrier concentration as functions of energy for (a) intrinsic, (b) n-type, and (c) p-type semiconductors at thermal equilibrium.

- Note: since the function f(E) is symmetrical about $E_F$, a large electron concentration implies a small hole concentration, and vice versa.
- In n-type material, the electron concentration in the conduction band increases as $E_F$ moves closer to $E_c$ ; thus, ( $E_c$ - $E_F$ ) gives a measure of n.
- Similarly, in p-type material, the hole concentration in the valence band increases as $E_F$ moves closer to $E_v$ ; thus, ( $E_v$ - $E_F$ ) gives a measure of p.

## Electron and Hole Concentrations at Equilibrium

- The F-D distribution function can be used to calculate the electron and hole concentrations in semiconductors, if the densities of available states in the conduction and valence bands are known.
- In equilibrium, the concentration of electrons in the conduction band can be given by

$$n_0 = \int_{E_0}^{\infty} f(E)N(E)dE$$

where N(E)dE is the density of available states/cm³ in the energy range dE.

- Note: the upper limit of is theoretically not proper, since the conduction band does not extend to infinite energies; however, since f(E) decreases rapidly with increasing E, the contribution to this integral for higher energies is negligible.
- Using the solution of *Schrodinger*'s wave equation under periodic boundary conditions, it can be shown that

$$N(E) = \frac{1}{2\pi^2}\left[\frac{2m^*}{(h/2\pi)^2}\right]^{3/2} E^{1/2}$$

- Thus, N(E) increases with E, however, f(E) decreases rapidly with E, thus, the product f(E) N(E) decreases rapidly with E, and very few electrons occupy states far above the conduction band edge, i.e., most electrons occupy a narrow energy band near the conduction band edge.
- Similarly, the probability of finding an empty state in the valence band [1 - f(E)] decreases rapidly below $E_V$, and most holes occupy states near the top of the valence band.
- Thus, a mathematical simplification can be made assuming that all available states in the conduction band can be represented by an effective density of states NC located at the conduction band edge $E_C$ and using Boltzmann approximation.

  Thus, $n_0 = N_c f(E_c) = N_c e^{-(E_c - E_F)/kT}$

  where . $N_c = 2\left(\frac{2\pi m_n^* kT}{h^2}\right)^{3/2}$

- Note: as ( $E_C$ - $E_F$ ) decreases, i.e., the Fermi level moves closer to the conduction band, the electron concentration increases.

  By similar arguments,

  $p_0 = N_v[1 - f(E_v)] = N_v e^{-(E_F - E_V)/kT}$

  where $N_V$ is the effective density of states located at the valence band edge $E_V$.

- Note: the only terms separating the expressions for $N_C$ and $N_v$ are the effective masses of electrons ( $m_n^x$ ) and holes ( $m_p^x$ ) respectively, and since , hence, $N_C$ $N_V$.
- Thus, as ( $E_F$ - $E_V$) decreases, i.e., the Fermi level moves closer to the valence band edge, and the hole concentration increases.
- These equations for $n_O$ and $p_O$ are valid in equilibrium, irrespective of the material being intrinsic or doped.

- For intrinsic material $E_F$ lies at an intrinsic level $E_i$ (very near the middle of the band gap), and the intrinsic electron and hole concentrations are given by

  $n_i = N_C e^{-(E_C - E_i)/kT}$ *and* $p_i = N_v e^{-(E_i - Ev)/kT}$

- Note: At equilibrium, the product $n_0 p_0$ is a constant for a particular material and temperature, even though the doping is varied,

  i.e., $n_0 p_0 = n_i^2 = N_C N_V e^{-E_g/kT}$

- This equation gives an expression for the intrinsic carrier concentration ni as a function of $N_C$, $N_V$, and temperature:

  $n_i = \sqrt{N_C N_V} e^{-E_g/kT}$

- These relations are extremely important, and are frequently used for calculations.
- Note: if $N_C$ were to be equal to $N_v$, then $E_i$ would have been exactly at mid gap (i.e., $E_c$ - $E_i$ = $E_i$ - $E_V$ = $E_g/2$).
- However, since $N_C$ $N_V$, $E_i$ is displaced slightly from mid gap (more for GaAs than that for Si).
- Alternate expressions for $n_0$ and $p_0$:

  $n_0 = N_i e^{(E_F - E_i)/kT}$ *and* $p_0 = N_i e^{(E_i - E_F)/kT}$

- Note: the electron concentration is equal to ni when $E_F$ is at $E_i$, and no increases exponentially as $E_F$ moves away from $E_i$ towards the conduction band.
- Similarly, the hole concentration $p_0$ varies from $n_i$ to larger values as $E_F$ moves from $E_i$ towards the valence band.

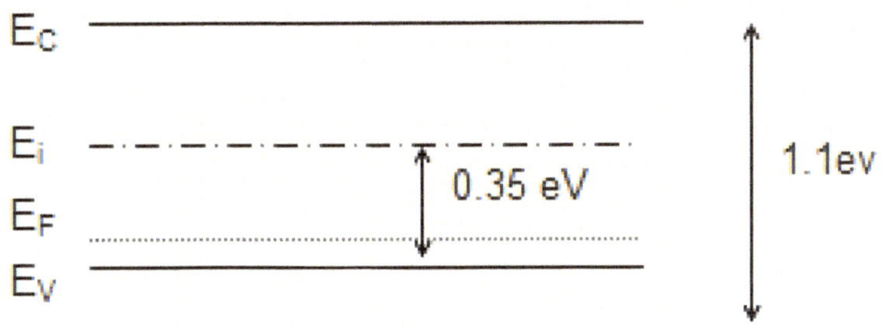


## Temperature Dependence of Carrier Concentrations

- The intrinsic carrier concentration has a strong temperature dependence, given by

  $$n_i(T) = 2\left(\frac{2\pi kT}{h^2}\right)^{3/2} (m_n^* m_p^*)^{3/4} e^{-E_g/2kT}$$

- Thus, explicitly, ni is proportional to T3/2 and to e 1/T, however, Eg also has a temperature dependence (decreasing with increasing temperature, since the interatomic spacing changes with temperature).

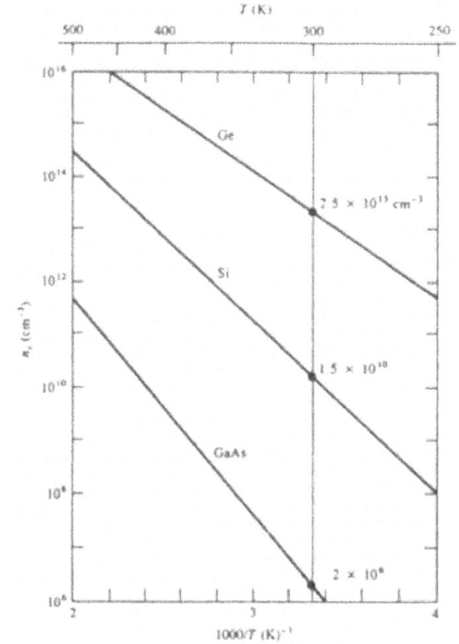


The intrinsic carrier concentration as a function of inverse temperature for Si, Ge, and GaAs.

- As $n_i$ changes with temperature, so do $n_0$ and $p_0$.
- With $n_i$ and T given, the unknowns are the carrier concentrations and the Fermi level position with respect to $E_i$ one of these quantities must be given in order to calculate the other.
- Example: Si doped with $10^{15} / cm^3$ donors ($N_d$).
- At very low temperature, negligible intrinsic EHPs exist, and all the donor electrons are bound to the donor atoms.
- As temperature is raised, these electrons are gradually donated to the conduction band, and at about 100 K (1000/T = 10), almost all these electrons are donated => this temperature range is called the ionization region.

Once all the donor atoms are ionized, the electron concentration $n_o \approx N_d$, since for each donor atom, one electron is obtained.

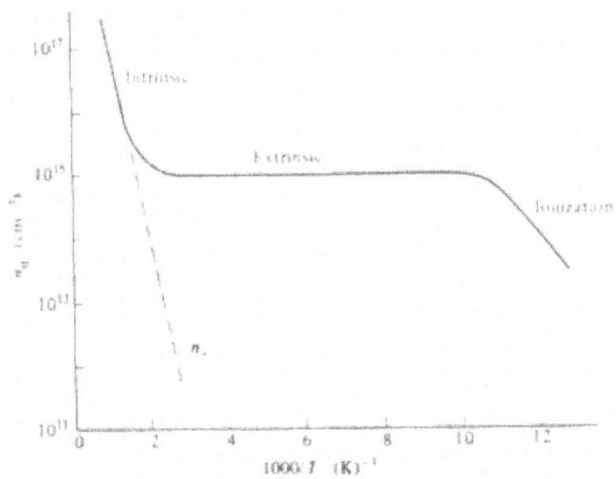


Variation of carrier concentration with inverse temperature clearly showing the three regions: ionization, extrinsic, and intrinsic.

- Thus, $n_0$ remains virtually constant with temperature for a wide range of temperature (called the extrinsic region), until the intrinsic carrier concentration ni starts to become comparable to $N_d$.
- For high temperatures, $n_i >> N_d$, and the material loses its extrinsic property (called the intrinsic region).
- Note: in the intrinsic region, the device loses its usefulness => determines the maximum operable temperature range.

## Compensation and Space Charge Neutrality

- Semiconductors can be doped with both donors ($N_d$) and acceptors ($N_A$) simultaneously.
- Assume a material doped with $N_d > N_A$ => predominantly n-type => $E_F$ lies above $E_i$ => acceptor level Ea completely full, however, with $E_F$ above $E_i$, the hole concentration cannot be equal to $N_A$.
- Mechanism:
  - Electrons are donated to the conduction band from the donor level $E_d$
  - An acceptor state gets filled by a valence band electron, thus creating a hole in the valence band.
  - An electron from the conduction band recombines with this hole.
  - Extending this logic, it is expected that the resultant concentration of electrons in the conduction band would be $N_d$ $N_A$ instead of $N_d$.
  - This process is called compensation.
- By compensation, an n-type material can be made intrinsic (by making $N_A = N_d$) or even p-type (for $N_A > N_d$).

  Note: a semiconductor is neutral to start with, and, even after doping, it remains neutral (since for all donated electrons, there are positively charged ions ($N_d^+$); and for all accepted electrons (or holes in the valence band), there are negatively charged ions ($N_a^-$).
- Therefore, the sum of positive charges must equal the sum of negative charges, and this governing relation,

  given by $p_0 + N_d^+ = n_0 + N_a^-$ is referred to as the equation for space charge neutrality.
- This equation, solved simultaneously with the law of mass action (given by $n_0 p_0 = n_i^2$) gives the information about the carrier concentrations.

  Note: for , $N_d^+ - N_a^- >> n_i^2 / n_0, n_0 \approx N_d^+ - N_a^-$, *and similarly, for* $N_a^+ - N_d^- >> n_i^2 / p_0, p_0 \approx N_a^+ - N_d^-$.

## Drift of Carriers in Electric and Magnetic Fields

- In addition to the knowledge of carrier concentrations, the collisions of the charge carriers with the lattice and with the impurity atoms (or ions) under electric and/or magnetic fields must be accounted for, in order to compute the current flow through the device.

- These processes will affect the ease (*mobility*) with which carriers move within a lattice.
- These collision and scattering processes depend on temperature, which affects the thermal motion of the lattice atoms and the velocity of the carriers.

## Conductivity and Mobility

- Even at thermal equilibrium, the carriers are in a constant motion within the lattice.
- At room temperature, the thermal motion of an individual electron may be visualized as random scattering from lattice atoms, impurities, other electrons, and defects.
- There is no net motion of the group of n $electrons / cm^3$ electrons/cm³ over any period of time, since the scattering is random, and there is no preferred direction of motion for the group of electrons and no net current flow.
- However, for an individual electron, this is not true the probability of an electron returning to its starting point after time t is negligibly small.
- Now, if an electric field $E_x$ is applied in the x-direction, each electron experiences a net force q $E_x$ from the field.
- This will create a net motion of group in the x-direction, even though the force may be insufficient to appreciably alter the random path of an individual electron.
- If $p_x$ is the x-component of the total momentum of the group, then the force of the field on the n $electrons / cm^3$ is

$$-nqE_x = \left.\frac{dp_x}{dt}\right|_{field}$$

  Note: this expression indicates a constant acceleration in the x-direction, which realistically cannot happen.

- In steady state, this acceleration is just balanced by the deceleration due to the collisions.
- Thus, while the steady field $E_x$ does produce a net momentum $p_{-x}$, for steady state current flow, the net rate of change of momentum must be zero when collisions are included.
- Note: the collision processes are totally random, thus, there is a constant probability of collision at any time for each electron.
- Consider a group of $N_0$ electrons at time t = 0, and define N(t) as the number of electrons that have not undergone a collision by time t

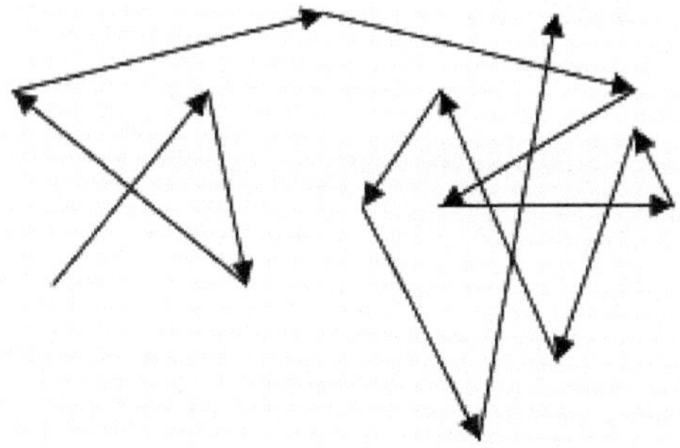

The random thermal motion of an individual electron, undergoing random scattering.

- The rate of decrease of N(t) at any time t is proportional to the number left unscattered at t, i.e.

  $$-\frac{dN(t)}{dt} = \frac{1}{\bar{t}} N(t)$$

  where $\bar{t}^{-1}$ is the constant of proportionality.

- The solution is an exponential function

  $$N(t) = N_0 e^{-t/\bar{t}}$$

- and $\bar{t}$ represents the mean time between scattering events, called the mean free time.
- The probability that any electron has a collision in time interval dt is dt/ $\bar{t}$ , thus, the differential change in $p_x$ due to collisions in time dt is

  $$dp_x = -p_x \frac{dt}{\bar{t}}$$

- Thus, the rate of change of $p_x$ due to the decelerating effect of collisions is

  $$\left.\frac{dp_x}{dt}\right|_{collisions} = -\frac{p_x}{\bar{t}}$$

- For steady state, the sum of acceleration and deceleration effects must be zero, thus,

  $$-\frac{p_x}{\bar{t}} - nqE_x = 0$$

- The average momentum per electron (averaged over the entire group of electrons) is

  $$\langle p_x \rangle = \frac{p_x}{n} = -q\bar{t}E_x$$

- Thus, as expected for steady state, the electrons would have on the average a constant net velocity in the -x-direction

  $$\langle v_x \rangle = \frac{\langle p_x \rangle}{m_n^*} = -\frac{q\bar{t}}{m_n^*}E_x$$

- This speed is referred to as the drift speed, and, in general, it is usually much smaller than the random speed due to thermal motion $v_{th}(=\sqrt{3kT / m_n^*})$.
- The current density resulting from this drift

$$J_x = -qn < v_x > = \frac{nq^2 \bar{t}}{m_n^*} E_x = \sigma E_x$$

- This is the familiar Ohm's law with $\sigma \equiv \frac{nq^2 \bar{t}}{m_n^*}$ being the conductivity of the sample, which can also be written as $\sigma = q\eta\mu_n$, with $\mu_n = \frac{q\bar{t}}{m_n^*}$ is defined as the *electron mobility* (in $cm^2 / V - \sec$), and it describes the ease with which electrons drift in the material.
- The mobility can also be expressed as the average drift velocity per unit electric field, thus $\mu_n = - < V_x > / E_x$ with the negative sign denoting a positive value for mobility since electrons drift opposite to the direction of the electric field.
- The total current density can be given by $J_x = q(\eta\mu_n + p\mu_p)E_x$ when both electrons and holes contribute to the current conduction; on the other hand, for predominantly n-type or p-type samples, respectively the first or the second term of the above equation dominates.

  Note: both electron and hole drift currents are in the same direction, since holes (with positive charges) move along the direction of the electric field, and electrons (with negative charges) drift opposite to the direction of the electric field.
- Since GaAs has a strong curvature of the E-k diagram at the bottom of the conduction band, the electron effective mass in GaAs is very small => the electron mobility in GaAs is very high since $\mu_n$ is inversely proportional to $m_n^*$.
- The other parameter in the mobility expression, i.e., $\bar{t}$ (the mean free time between collisions) is a function of temperature and the impurity concentration in the semiconductor.
- For a uniformly doped semiconductor bar of length L, width w, and thickness t, the resistance R of the bar can be given by $L / (Wt\sigma) = \rho L / (wt)$ where $\rho$ is the resistivity.

## Effects of Temperature and Doping on Mobility

- The two main scattering events that influence electron and hole motion (and, thus, mobility) are the lattice scattering and the *impurity scattering*.
- All lattice atoms vibrate due to temperature and can scatter carriers due to collisions.
- These collective vibrations are called phonons, thus *lattice scattering* is also known as *phonon scattering*.
- With increasing temperature, lattice vibrations increase, and the mean free time between collisions decreases => mobility decreases (typical dependence $\sim T^{-3/2}$).
- Scattering from crystal defects and ionized impurities dominate at low temperatures.

- Since carriers moving with low velocity (at low temperature) can get scattered more easily by ionized impurities, this kind of scattering causes a decrease in carrier mobility with decreasing temperature (typical dependence $\sim T^{3/2}$).
- Note: the scattering probability is inversely proportional to the mean free time (and to mobility), hence, the mobilities due to two or more scattering events add inversely:

$$\frac{1}{\mu} = \frac{1}{\mu_1} + \frac{1}{\mu_2} + \ldots$$

- Thus, the mechanism causing the lowest mobility value dominates.
- Mobility also decreases with increasing doping, since the ionized impurities scatter carriers more (e.g., $\mu_n$ for intrinsic Si is 1350 $cm^2 / V-$sec at 300 K, whereas with a donor doping of $10^{17} / cm^3$ , n drops to 700 $cm^2 / V-$sec ).

## High Field Effects

- For small electric fields, the drift current increases linearly with the electric field, since $\sigma$ is a constant.
- However, for large electric fields (typically > $10^3$ $V / cm$ ), the current starts to show a sublinear dependence on the electric field and eventually saturates for very high fields.
- Thus, becomes a function of the electric field, and this is known as the hot carrier effect, when the carrier drift velocity becomes comparable to its thermal velocity.
- The maximum carrier drift velocity is limited to its mean thermal velocity (typically $10^7$ $cm /$ sec ), beyond which the added energy imparted by the electric field is absorbed by the lattice (thus generating heat) instead of a corresponding increase in the drift velocity.

## The Hall Effect

- An extremely important measurement procedure for determining the majority carrier concentration and mobility.

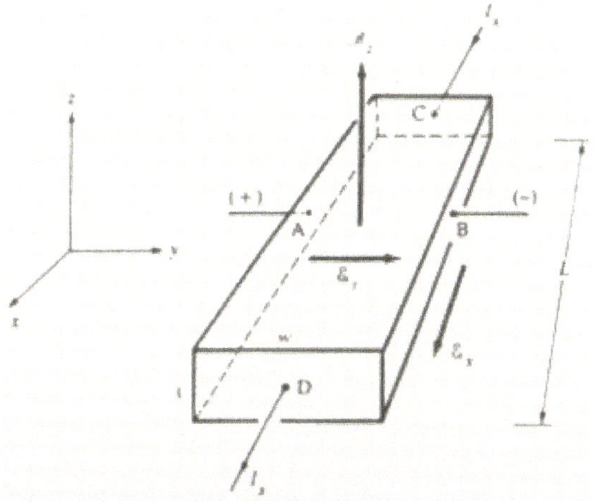


The experimental setup for the Hall Effect measurement.

- If a magnetic field is applied perpendicular to the direction of carrier flow, the path of the carriers get deflected due to the Lorentz force experienced by the carriers, which can be given by
- F = q(E + v x B)
- Thus, the holes will get deflected towards the -y-direction, and establish an electric field along the y-direction, such that in steady state $E_V = v_x B_z$
- The establishment of this electric field is known as the *Hall effect*, and the resulting voltage $V_{AB} = E_{yW}$ is called the *Hall voltage.*
- Using the expression for the drift current, $E_V = R_H J_X B_Z$, *where* $R_H(=[qp_0]^{-1})$ is called the Hall coefficient.
- A measurement of the Hall voltage along with the information for magnetic field and current density gives the majority carrier concentration $p_0(= I_X B_Z / [qtV_{AB}])$
- Also, the majority carrier mobility $\mu_0(= R_H / p)$ can be obtained from a measurement of the resistivity $\rho(=[V_{CD} / I_X]/[L/(wt)])$
- This experiment can be performed to obtain the variation of majority carrier concentration and mobility as a function of temperature.
- For n-type samples, the Hall voltage and the Hall coefficient are negative => a common diagnostic tool for obtaining the sample type.
- Note: caution should be exercised for near intrinsic samples.

## Equilibrium Condition

- In equilibrium, there is no external excitation except a constant temperature, no net transfer of energy, no net carrier motion, and no net current transport.
- An important condition for equilibrium is that no discontinuity or gradient can arise in the equilibrium Fermi level EF.
- Assume two materials 1 and 2 (e.g., n- and p-type regions, dissimilar semiconductors, metal and semiconductor, two adjacent regions in a nonuniformly doped semiconductor) in intimate contact such that electron can move between them.
- Assume materials 1 and 2 have densities of state N1(E) and N2(E), and F-D distribution functions f1(E) and f2(E) respectively at any energy E.

$$\textit{The Sample is } p-\textit{type with } p_0 \gg n_0, \textit{and the mobility } \mu_\rho \approx 400\, cm^2 / V - \text{sec}.$$

$$\textit{Thus, the resistivity } p = (\sigma)^{-1} = (q\mu_\rho P_0)^{-1} = 1.5625\ \Omega - cm.$$

$$\textit{The Hall coefficient } R_H = +(qp_0)^{-1} = +625\ cm^3 / C\ (\text{note } \textit{the unit of } R_H).$$

$$\textit{Note, alternately } R_H = \rho\mu_\rho = 625\, cm^3 / C.$$

$$\textit{The Hall Voltage } V_{AB} = (I_X B_Z R_H)/t = (2\times10^{-3})(5\times10^{-5})(625)/(150\times10^{-4}) = 4.17 mV.$$

- The rate of electron motion from 1 to 2 can be given byrate from 1 to 2 N1(E)f1(E) . N2(E)[1 f2(E)] (2.30)and the rate of electron motion from 2 to 1 can be given byrate from 2 to 1 N2(E) f2(E) . N1(E)[1 f1(E)] (2.31)" At equilibrium, these two rates must be equal, which gives f1(E) = f2(E) => EF1 = EF2 => dEF/dx = 0; thus, the Fermi level is constant at equilibrium, or, in other words, there cannot be any discontinuity or gradient in the Fermi level at equilibrium.

# Charge Carrier

In physics, a charge carrier is a particle free to move, carrying an electric charge, especially the particles that carry electric charges in electrical conductors. Examples are electrons, ions and holes. In a conducting medium, an electric field can exert force on these free particles, causing a net motion of the particles through the medium; this is what constitutes an electric current. In different conducting media, different particles serve to carry charge:

- In metals, the charge carriers are electrons. One or two of the valence electrons from each atom is able to move about freely within the crystal structure of the metal. The free electrons are referred to as conduction electrons, and the cloud of free electrons is called a Fermi gas.
- In electrolytes, such as salt water, the charge carriers are ions, which are atoms or molecules that have gained or lost electrons so they are electrically charged. Atoms that have gained electrons so they are negatively charged are called anions, atoms that have lost electrons so they are positively charged are called cations. Cations and anions of the dissociated liquid also serve as charge carriers in melted ionic solids. Proton conductors are electrolytic conductors employing positive hydrogen ions as carriers.
- In a plasma, an electrically charged gas which is found in electric arcs through air, neon signs, and the sun and stars, the electrons and cations of ionized gas act as charge carriers.
- In a vacuum, free electrons can act as charge carriers. These are sometimes called cathode rays. In a vacuum tube, the mobile electron cloud is generated by a heated metal cathode, by a process called thermionic emission.
- In semiconductors (the material used to make electronic components like transistors and integrated circuits), in addition to electrons, the travelling vacancies in the valence-band electron population (called "holes"), act as mobile positive charges and are treated as charge carriers. Electrons and holes are the charge carriers in semiconductors.

It can be seen that in some conductors, such as ionic solutions and plasmas, there are both positive and negative charge carriers, so an electric current in them consists of the two polarities of carrier moving in opposite directions. In other conductors, such as metals, there are only charge carriers of one polarity, so an electric current in them just consists of charge carriers moving in one direction.

## Charge Carriers in Semiconductors

There are two recognized types of charge carriers in semiconductors. One is electrons, which carry a negative electric charge. In addition, it is convenient to treat the traveling vacancies in the valence band electron population (holes) as the second type of charge carrier, which carry a positive charge equal in magnitude to that of an electron.

## Carrier Generation and Recombination

When an electron meets with a hole, they recombine and these free carriers effectively vanish. The energy released can be either thermal, heating up the semiconductor (*thermal recombination*, one of the sources of waste heat in semiconductors), or released as photons (*optical recombination*, used in LEDs and semiconductor lasers). The recombination means an electron which has been excited from the valence band to the conduction band falls back to the empty state in the valence band, known as the holes. The holes are the empty state created in the valence band when an electron gets excited after getting some energy to overpass the energy gap.

## Majority and Minority Carriers

The more abundant charge carriers are called majority carriers, which are primarily responsible for current transport in a piece of semiconductor. In n-type semiconductors they are electrons, while in p-type semiconductors they are holes. The less abundant charge carriers are called minority carriers; in n-type semiconductors they are holes, while in p-type semiconductors they are electrons.

In an intrinsic semiconductor, which does not contain any impurity, the concentrations of both types of carriers are ideally equal. If an intrinsic semiconductor is doped with a donor impurity then the majority carriers are electrons; if the semiconductor is doped with an acceptor impurity then the majority carriers are holes.

Minority carriers play an important role in bipolar transistors and solar cells. Their role in field-effect transistors (FETs) is a bit more complex: for example, a MOSFET has both p-type and n-type regions. The transistor action involves the majority carriers of the source and drain regions, but these carriers traverse the body of the opposite type, where they are minority carriers. However, the traversing carriers hugely outnumber their opposite type in the transfer region (in fact, the opposite type carriers are removed by an applied electric field that creates an inversion layer), so conventionally the source and drain designation for the carriers is adopted, and FETs are called "majority carrier" devices.

## Free Carrier Concentration

*Free carrier concentration* is the concentration of free carriers in a doped semiconductor. It is similar to the carrier concentration in a metal and for the purposes of calculating currents or drift velocities can be used in the same way. Free carriers are electrons (or holes) which have been introduced directly into the conduction band (or valence band) by doping and are not promoted thermally. For this reason electrons (holes) will not act as double carriers by leaving behind holes (electrons) in the other band. In other words, charge carriers are particles/electrons that are free to move (carry the charge). They get charged at the battery/cell, and then flow throughout the circuit, giving energy to any components in the way, then when out of charge and towards the end of the circuit, they get re-charged by the battery/cell.

## Carrier Generation and Recombination

In the solid-state physics of semiconductors, carrier generation and recombination are processes by which mobile charge carriers (electrons and electron holes) are created and eliminated. Carrier generation and recombination processes are fundamental to the operation of many optoelectronic semiconductor devices, such as photodiodes, LEDs and laser diodes. They are also critical to a full analysis of p-n junction devices such as bipolar junction transistors and p-n junction diodes.

The electron–hole pair is the fundamental unit of generation and recombination, corresponding to an electron transitioning between the valence band and the conduction band where generation of electron is a transition from the valence band to the conduction band and recombination leads to a reverse transition.

## Band Structure

Like other solids, semiconductor materials have an electronic band structure determined by the crystal properties of the material. The actual energy distribution among the electrons is described by the Fermi level and the temperature of the electrons. At absolute zero temperature, all of the electrons have energy below the Fermi level; but at non-zero temperatures the energy levels are filled following a Boltzmann distribution.

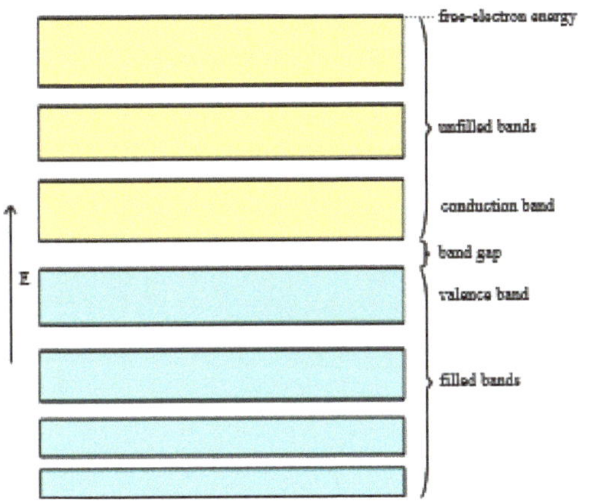

Electronic band structure of a semiconductor material.

In semiconductors the Fermi level lies in the middle of a *forbidden band* or band gap between two *allowed bands* called the *valence band* and the *conduction band*. The valence band, immediately below the forbidden band, is normally very nearly completely occupied. The conduction band, above the Fermi level, is normally nearly completely empty. Because the valence band is so nearly full, its electrons are not mobile, and cannot flow as electric current.

However, if an electron in the valence band acquires enough energy to reach the conduction band, it can flow freely among the nearly empty conduction band energy states. Furthermore, it will also leave behind an electron hole that can flow as current exactly like a physical charged particle. Carrier *generation* describes processes by which electrons gain energy and move from the valence band to the conduction band, producing two mobile carriers; while *recombination* describes processes by which a conduction band electron loses energy and re-occupies the energy state of an electron hole in the valence band.

In a material at thermal equilibrium generation and recombination are balanced, so that the net charge carrier density remains constant. The equilibrium carrier density that results from the balance of these interactions is predicted by thermodynamics. The resulting probability of occupation of energy states in each energy band is given by Fermi–Dirac statistics.

## Generation and Recombination Processes

Carrier generation and recombination occur when an electron makes transition from the valence band to conduction band in a semiconductor, as a result of interaction with other electrons, holes, photons, or the vibrating crystal lattice itself. These processes must conserve both quantized energy and momentum, and the vibrating lattice plays a large role in conserving momentum as photons carry very little momentum in relation to their energy.

Recombination and generation are always happening in semiconductors, both optically and thermally, and their rates are in balance at equilibrium. The product of the electron and hole densities ($n$ and $p$) is a constant ($n_o p_o = n_i^2$) at equilibrium, maintained by recombination and generation occurring at equal rates. When there is a surplus of carriers (i.e., $np > n_i^2$), the rate of recombination becomes greater than the rate of generation, driving the system back towards equilibrium. Likewise, when there is a deficit of carriers (i.e., $np < n_i^2$), the generation rate becomes greater than the recombination rate, again driving the system back towards equilibrium. As the electron moves from one energy band to another, the energy and momentum that it has lost or gained must go to or come from the other particles involved in the process (e.g. photons, electron, or the system of vibrating lattice atoms). The following models are used to describe generation and recombination, depending on which particles are involved in the process.

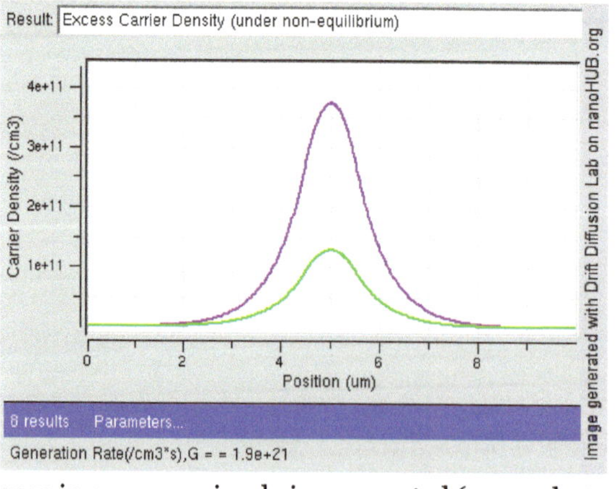


The following image shows change in excess carriers being generated (green:electrons and purple:holes) with increasing light intensity (Generation rate /cm3) at the center of an intrinsic semiconductor bar. Electrons have higher diffusion constant than holes leading to fewer excess electrons at the center as compared to holes.

## Shockley–Read–Hall (SRH) Process

In *Shockley-Read-Hall recombination*, also called *trap-assisted recombination*, the electron in transition between bands passes through a new energy state (localized state) created within the band gap by an impurity in the lattice; such energy states are called deep-level traps. The localized state can absorb differences in momentum between the carriers, and so this process is the dominant generation and recombination process in silicon and other indirect bandgap materials. It can also dominate in direct bandgap materials under conditions of very low carrier densities (very low level injection). The energy is exchanged in the form of lattice vibration, a phonon exchanging thermal energy with the material. The process is named after William Shockley, William Thornton Read and Robert N. Hall.

### Radiative Recombination

During *radiative recombination*, a form of spontaneous emission, a photon is emitted with the wavelength corresponding to the energy released. This effect is the basis of LEDs. Because the photon carries relatively little momentum, radiative recombination is significant only in direct bandgap materials.

When photons are present in the material, they can either be absorbed, generating a pair of free carriers, or they can *stimulate* a recombination event, resulting in a generated photon with similar properties to the one responsible for the event. Absorption is the active process in photodiodes, solar cells, and other semiconductor photodetectors, while stimulated emission is responsible for laser action in laser diodes.

### Auger Recombination

In *Auger recombination* the energy is given to a third carrier, which is excited to a higher energy level without moving to another energy band. After the interaction, the third carrier normally loses its excess energy to thermal vibrations. Since this process is a three-particle interaction, it is normally only significant in non-equilibrium conditions when the carrier density is very high. The Auger effect process is not easily produced, because the third particle would have to begin the process in the unstable high-energy state.

The Auger recombination can be calculated from the equation :

$$U_{Aug} = \Gamma_n n(np - n_i^2) + \Gamma_p p(np - n_i^2)$$

### Auger Recombination in LEDs

The mechanism causing LED efficiency droop was identified in 2007 as Auger recombination, which met with a mixed reaction. In 2013, an experimental study claimed to have identified Auger recombination as the cause of efficiency droop. However, it remains disputed whether the amount of Auger loss found in this study is sufficient to explain the droop. Other frequently quoted evidence against Auger as the main droop causing mechanism is the low-temperature dependence of this mechanism which is opposite to that found for the drop.

## Excess Carriers In Semiconductors

- Excess carriers, essential for device operation, are created by optical excitation, electron bombardment, or injected across a forward-biased p-n junction.
- These excess carriers can dominate the conduction process in semiconductor materials.

## Optical Absorption

- This includes photons in the optical range as well as those in the infrared region.
- Photons of various wavelengths (frequencies) are directed at the sample, and their relative transmission is measured.
- Note: photons having energies greater than the band gap energy are absorbed (the sample behaves opaque for this kind of illumination), whereas those having energies less than the band gap energy are transmitted (the sample behaves transparent), this experiment gives an accurate measure of the band gap energy.
- When photons having energies h $E_g$ are absorbed, they create EHPs and the probability of this absorption is very high, since there are lots of electrons in the valence band and lots of empty states in the conduction band.
- Electrons excited to EC may initially have energies much higher than EC, however, they lose this excess energy due to scattering with the lattice until their equilibrium energy becomes equal to EC.
- Note: these EHPs are called excess carriers, since they are out of balance, and, thus, would eventually recombine.
- However, while these excess carriers remain in the respective bands, they can contribute to the current conduction.
- The transmitted intensity It of a beam of photons of wavelength through a sample of thickness t can be given by

  $I_t = I_0 e^{-\alpha t}$

  where $\alpha(cm^{-1})$ is called the absorption oefficient, and varies with materials and photon wavelength

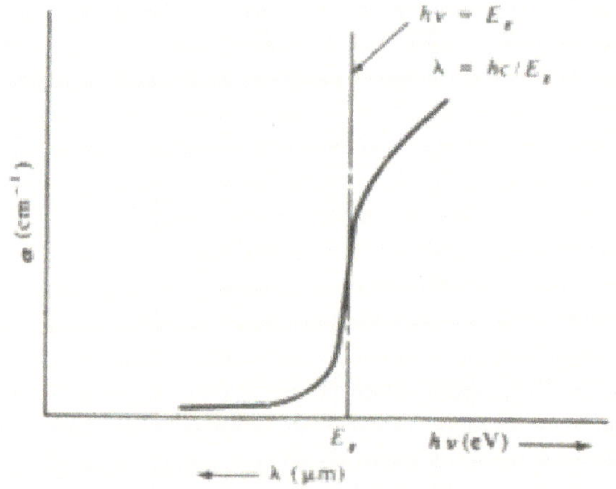


The variation of the absorption coefficient as a function of the wavelength of the incident light.

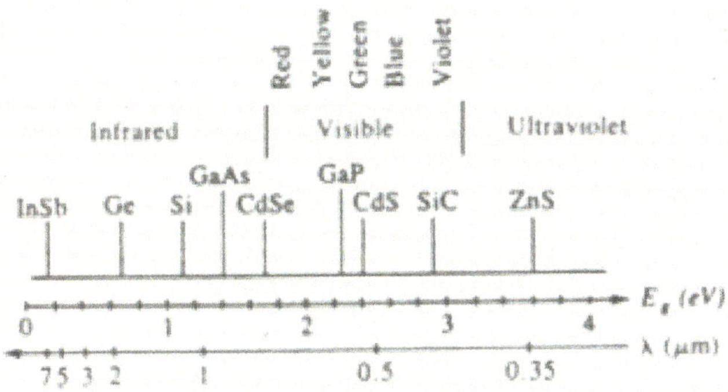


Band gaps of some common semiconductors relative to the optical spectrum.

- Absorption cutoff occurs at $\lambda_0 = 1.24 / E_g (E_g \, in \, eV)$
- GaAs, Si, Ge, and InSb band gaps are such that $\lambda_0$ c occurs beyond the visible region (in the infrared), whereas GaP and CdS have band gaps with $\lambda_0$ c falling within the visible range.

## Luminescence

- When recombination occurs between a conduction band electron and a valence band hole, the energy released can be given off in the form of light (luminescence). Direct band-to-band recombination in direct band gap semiconductors have a much higher probability of light emission as compared to those in indirect materials.
- Broadly divided into three categories:
- Photoluminescence: if the recombining carriers were caused by optical excitation.
- Cathodoluminescence: if the recombining carriers were caused by high energy electron bombardment.
- Electroluminescence: if the recombining carriers were caused by injection of excess carriers (by forward biasing a p-n junction, for example).

## Photoluminescence

- For steady state excitation, the recombination rate and the generation rate for EHPs are equal, and one photon is emitted for each photon absorbed.
- Direct band-to-band recombination is a fast process with typical lifetime of excess carriers 10 8 sec => known as fluorescence (example: fluorescent lamp).
- In some indirect materials, the trap states within the band gap captures carriers, and slows down the recombination process, thus, emission continues for seconds or minutes after the excitation is removed => known as phosphorescence and the materials are known as phosphors.
- The trap states can hold the carriers for indefinite times, and the carriers can either get reexcited to the conduction band or fall to the valence band (and, thus, recombine) => this creates the delay between excitation and recombination.

## Cathodoluminescence

- Best example: cathode ray tube (CRT) basis of television sets, oscilloscopes, and other display systems.
- Electrons emitted from the heated cathode are accelerated towards the anode by high field, deflected by electric or magnetic fields by the horizontal and vertical plates, and made to hit the screen (coated with a phosphor) at desired locations.
- When electrons hit these phosphors, the energy of the electrons gets transferred to the electrons of the phosphors, and they get excited to higher states, and eventually fall to the ground state, thus causing recombination and light emission.
- Three phosphor dots are used for each pixel, capable of transmitting three primary colors: red, green, and blue (RGB) thus by varying the intensity and position of the electron beam, a wide range of colors and picture can be attained.

## Electroluminescence

- Best examples: LEDs and LASERs, where carriers are injected across a forward biased p-n junction and are made to recombine (either naturally or by carrier confinement) => called injection electroluminescence.

## Carrier Lifetime and Photoconductivity

- Photoconductivity: increase in the conductivity of a sample due to excess carriers created by optical excitation.
- With excitation turned off, the photoconductivity decreases to zero since all excess carriers eventually recombine.

## Direct Recombination of Electrons and Holes

- Direct recombination occurs spontaneously, i.e., the probability that an electron and a hole will recombine is constant in time, which leads to an exponential solution for the decay of the excess carriers.
- The net rate of change in the conduction band electron concentration at any time t

  $$\frac{dn(t)}{dt} = \alpha_r n_i^2 - \alpha_r n(t)p(t)$$

  where the first term is the generation rate and the second term is the recombination rate.
- Let excess EHPs n and p (with n = p, since they are created in pairs) are created at t = 0 by a short flash of light.
- Define n(t) and p(t) (again n = p) as the instantaneous excess carrier concentrations and n and p for their values at t = 0.

- Note: n(t) = no + n(t), and p(t) = po + p(t).
- Thus,

$$\frac{d\delta n(t)}{dt} = \alpha_r p_0 \delta n(t)$$

$$\delta n(t) = \Delta n e^{-t/\tau}$$

where $\tau_n (= [\alpha_r p_0]^{-1})$ is called the minority carrier recombination lifetime or simply the minority carrier lifetime, and it determines the rate at which the minority carriers recombine with time.

- Similarly, excess holes in an n-type material recombine with a rate $\tau_n (= [\alpha_r p_0]^{-1})$
- Note: for direct recombination, the excess majority carriers (which is equal to the excess minority carriers) decay at exactly the same rate as the minority carriers, however, there is a large percentage change in the minority carriers as compared to the majority carriers.
- A more general expression for carrier lifetime for near not sufficiently extrinsic samples is $\tau = 1/[\alpha_r (n_0 + p_0)]$.

## Indirect Recombination: Trapping

- In indirect materials, the probability of direct band-to-band recombination is very small recombination in these materials proceed through the assistance of recombination (or trapping) centers located within the band gap, which trap carriers of one type followed by trapping carriers of the opposite type, thus annihilating the pairs.

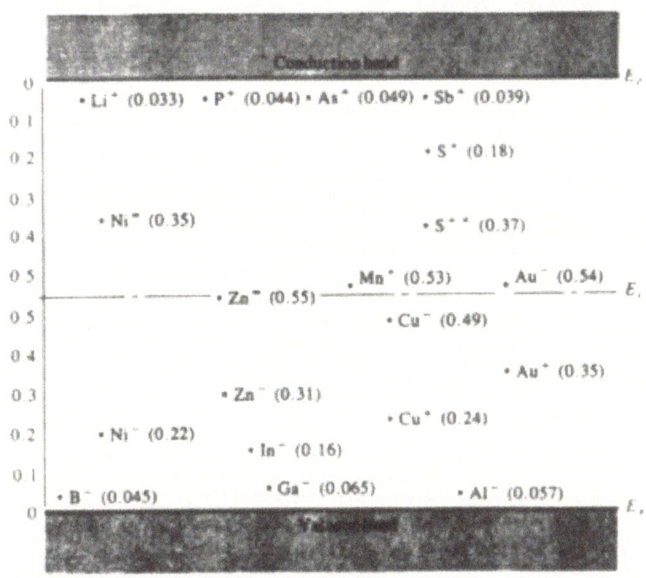


Energy levels of common impurities in silicon, measured from the closest band edge.

- The resulting energy loss is often in the form of heat given to the lattice (instead of light emission), and, thus, these materials are not well suited for optoelectronic applications.

- There are four probabilities associated with a recombination center:

  (a) hole capture: when an electron from the recombination center falls to the valence band,

  (b) hole emission: when an electron makes a transition from the valence band to the recombination center,

  (c) electron capture: when an electron falls from the conduction band to the recombination center, and

  (d) electron emission: when an electron makes a transition from the recombination center to the conduction band.

- Each of these processes has their own probabilities and time constants, and the resulting analysis is significantly complicated.
- This theory of recombination is known as the SHR (Shockley-Hall-Read) theory of recombination.
- If process (a) follows process (c) or vice versa, recombination takes place, whereas, if process (b) follows process (a) (or vice versa) or process (c) follows process (d) (or vice versa), it is known as reemission, and the recombination center behaves like a trapping center.
- Generally, centers that are located towards the middle of the band gap (e.g., Au in Si) behave like recombination centers, whereas centers located closer to the conduction or valence band behave as traps, for obvious reasons.
- Alternate definition: in a center located within the band gap, if after capturing one type of carrier, the most probable next event is the capture of opposite type of carrier, then it is a recombination center, however, if the most probable next event is reexcitation, then it is a trap.
- The recombination can be slow or fast, depending on the amount of time the carrier spends in the center before the capture of the opposite type of carrier happens, thus, computation of lifetime for this kind of indirect recombination is sufficiently complicated.
- The decay of excess carriers can be measured by a typical photoconductive decay measurement, where light shining on a sample is suddenly switched off, and the resulting decay of current passing through the sample is measured, the rate of decay of this current gives the excess carrier lifetime.

## Auger Recombination

- Note: the lifetime is proportional to the inverse of the doping concentration.
- However, at relatively high doping levels, the lifetime decreases at a faster rate with an increase in the doping concentration.
- This is because a different recombination mechanism, called the Auger recombination becomes dominant at high doping levels.

- In this recombination mechanism, the electron and hole recombine without involving trap levels, and the released energy (of the order of the energy gap) is transferred to another carrier (a hole in p-type material and an electron in n-type material).
- This process is somewhat the inverse of the impact ionization process, in which an energetic carrier causes EHP generation.
- Since two electrons (in n-type material) or two holes (in p-type material) are involved in this process, it is highly unlikely except in heavily doped material.
- The recombination lifetime associated with the Auger recombination process is inversely proportional to the square of the majority carrier concentration, i.e., for p-type material, $\tau_n \approx 1/(G_p N_A^2)$ , and for n-type material, $\tau_n \approx 1/(G_p N_D^2)$ , where Gp and Gn are coefficients with values of $9.9 \times 10^{-32}\, cm^6 / \sec$ and $2.28 \times 10^{-31}\, cm^6 / \sec$ .

## Surface Recombination

- It is obvious that near the surface of any semiconductor device, the carrier recombination rate should be very high, due to extra defects and traps at the surface.
- Thus, the diffusion flux of minority carriers at the surface is determined by the surface recombination processes.

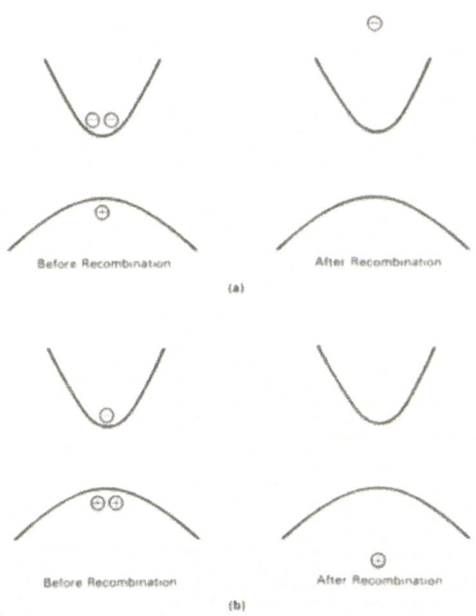


Auger recombination: (a) n-type sample, and (b) p-type sample.

- For example, when minority carriers are holes, this surface recombination can be described by:

$$D_p \left. \frac{\partial p_n}{\partial x} \right|_{x=0} = -S_p [p_n(x=0) - p_{n0}]$$

- where x = 0 corresponds to the surface of the sample, and $S_p = \sigma_p V_{thp} N_{st}$ is the surface recombination rate, with $\sigma_p$ being the capture cross-section for holes, $V_{thp}$ is the thermal velocity for the holes, and $N_{st}$ is the surface density of the surface states.

- The capture cross-section (typically $10^{-15}\ cm^2$) describes the effectiveness of the localized state in capturing a carrier.
- The product $V_{thp}$ may be visualized as the volume swept out per unit time by a particle with cross-section $\sigma$.
- If the surface state lies within this volume around the carrier, then the carrier gets captured by the surface state.
- Note: the dimension of S is cm/sec, and, consequently, it is termed as the surface recombination velocity, even though it has nothing to do with actual velocity.

## Steady State Carrier Generation: Quasi-Fermi Levels

- For any temperature T, there is a thermal generation rate g(T) balanced by a recombination rate r(T).
- Now, if a steady light is shone on the sample, an optical generation rate $g_{op}$ will be added to g(T), and the carrier concentrations n and p would increase to their new steady state values.
- Generation/recombination rate balance equation:

  $$g(T)+g_{op}=\alpha_r np=\alpha_r(n_0+\delta n)(p_0+\delta p)$$

- For steady state recombination and no trapping, $\delta n=\delta p$; and, under low level injection approximation

  $$g_{op}=\alpha_r(n_0+p_0)\delta n=\delta n/\tau_n$$

- Thus, the excess carrier concentrations can be given by $\delta n=\delta p=g_{op}\tau_n$
- In general, when trapping is present, $\tau_n\neq\tau_p$, and $\delta n=g_{op}\tau_n\ and\ \delta p=g_{op}\tau_{p.}$ and $\delta p=g_{op}\tau_{p.}$.
- Note: when excess carriers are present, $np\neq n_i^2$.
- When excess carriers are present, the equilibrium Fermi level $E_F$ is no more meaningful; instead, the carrier concentrations are defined in terms of quasi-Fermi levels (also referred to as Imref, which is Fermi spelled backwards) $F_n\ and\ F_p$ as

  $$n=n_ie^{(F_0-E_1)/kT}\quad and\quad p=n_ie^{(E_1-F_0)/kT}$$

- Imref for the minority carriers deviates significantly from the equilibrium Fermi level, whereas, for majority carriers, the Imref stays very close to the equilibrium Fermi level, and the separation between these two Imrefs is a measure of how far the system is from equilibrium.
- With concentrations varying with position, the Imrefs would also vary with position, thus

drawing Imrefs in band diagrams clearly shows the positional variations of the carrier concentrations.

## Photoconductive Devices

- Devices which change their resistance while exposed to light.
- Examples: automatic night light controllers, exposure meters in cameras, moving-object counters, burglar alarms, detectors in fiber optic communication systems, etc.
- Considerations in choosing a photoconductor for a given application: sensitive wavelength range, time response, and optical sensitivity (responsivity) of the material.
- The photoconductivity change while exposed to light is

  $$\Delta\sigma = qg_{op}(\tau_n\mu_n + \tau_p\mu_p)$$

- Obvious that for large changes in $\Delta\sigma$, the carrier mobility and lifetime should be high (e.g., in $InSb, \mu_n \approx 10^5\ cm^2/V-\sec$, and could be used as infrared detector with high sensitivity).
- Time response is limited by recombination times, degree of carrier trapping, and time required for the carriers to drift through the device in an electric field.
- Dark resistance (i.e., the resistance of the device without any illumination) should be as small as possible.
- Generally, all these requirements cannot be satisfied simultaneously, and some kind of optimization is required.

## Diffusion of Carriers

- When excess carriers are created in a semiconductor and their concentrations vary with position, then there is a net carrier motion from regions of higher concentration to regions of lower concentration.
- This type of motion is called the diffusion, and it is an important charge transport mechanism in semiconductors.
- Diffusion and drift are the two main current transport mechanisms.

## Diffusion processes

- Natural result of the random motion of individual electrons.
- Electrons move randomly and experience collisions, on the average, after each mean free time .
- Since the motion is truly random, an electron has equal probability of moving into or out of a volume through a boundary.

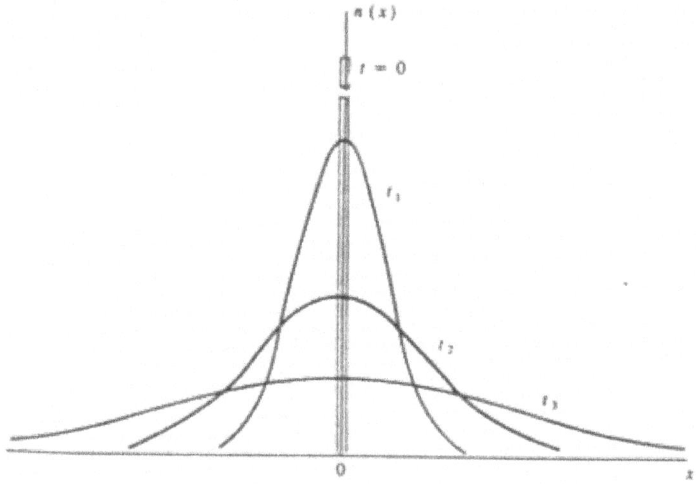


Spreading of a narrow pulse of electrons created at x = 0 at t = 0 with time $(t_3 > t_2 > t_1)$.

- A pulse of excess electrons injected at x = 0 at time t = 0 will spread out in time due to diffusion, and eventually n(x) becomes a constant, when no more net motion takes place.

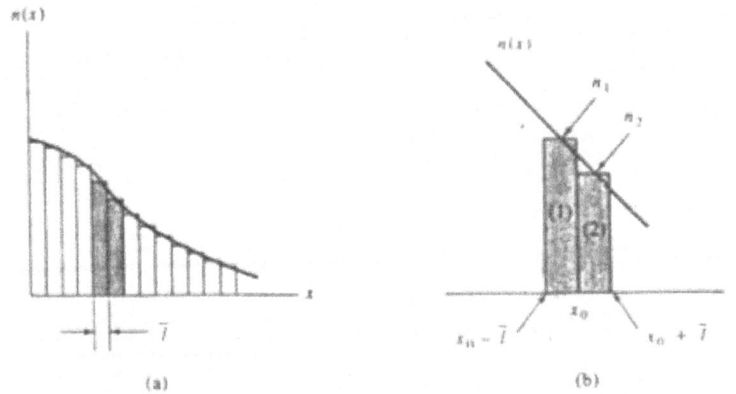


An arbitrary electron distribution along the x-direction: (a) each segment divided into lengths equal to the mean free path $l$, and (b) expanded view of two segments centered at $x_0$

- Consider any arbitrary distribution n(x), with x divided into segments $\bar{1}$ (mean free path) wide, with n(x) evaluated at the center of each segment.

- In $\bar{t}$, half of the electrons in segment (1) to the left of $x_0$ would move into segment (2), and in the same time, half of the electrons in segment (2) to the right of $x_0$ would move into segment (1).

- Therefore, the net number of electrons moving from segment (1) to segment (2) through $x_0$ within a mean free time $\bar{t} = (n_1 - n_2)\bar{1}A/2$, where A is the area perpendicular to x.

- Thus, the electron flux density in the +x-direction

$$\phi_n(x_o) = \frac{i}{2\bar{t}}(n_1 - n_2)$$

- Note: is a small differential length, thus,

$$n_1 - n_2 = \frac{n(x) - n(x+\Delta x)}{\Delta x} i$$

where x is taken at the center of segment (1) and x = $\bar{1}$.

- In the limit of small $\Delta x$ (i.e., small mean free path $\bar{1}$ between collisions)

$$\phi(x) = \frac{\bar{1}^2}{2\bar{t}} \lim_{dx \to 0} \frac{n(x) - n(x+\Delta x)}{\Delta x} = -\frac{\bar{1}^2}{2\bar{t}} \frac{dn(x)}{dx}$$

- The quantity $\bar{1}^2 / 2\bar{t}$ is called the electron diffusion coefficient $D_n (cm^2 / \sec)$.
- The minus sign in the expression for $\phi_n(x)$ implies that the diffusion proceeds from higher electron concentration to lower electron concentration.
- Similarly, holes diffuse from a region of higher concentration to a region of lower concentration with a diffusion coefficient $D_p$
- Thus,

  $$\phi_n(x) = -D_n \frac{dn(x)}{dx} \quad and \ \phi_p(x) = -D_p \frac{dp(x)}{dx}$$

  and the diffusion current density

  $$J_n(diff.) = +qD_n \frac{dn(x)}{dx} \quad and \quad J_p(diff.) = -qD_p \frac{dp(x)}{dx}$$

- Note: electrons and holes move together in a carrier gradient, however, the resulting currents are in opposite directions because of the opposite charges of the particles.

## Diffusion and Drift of Carriers: Built-in Fields

- The total current density can thus be written as

  $$J_n(x) = \underbrace{q\mu_n n(x)E(x)}_{drif} + \underbrace{qD_n \frac{dn(x)}{dx}}_{diffusion}$$

  $$J_p(x) = q\mu_p p(x)E(x) + qD_p \frac{dp(x)}{dx}$$

  and the total current density is

  $$J(x) = J_p(x) + J_n(x)$$

- The total current may be due primarily to one of these two components, depending upon the carrier concentrations, their gradients, and the electric field.
- Thus, minority carriers can contribute to current conduction significantly through diffu-

sion, even though they contribute very little to the drift term (due to their low concentrations).

- Since electrons drift opposite to the direction of the electric field (due to their negative charge), their potential energy increases in the direction of the electric field.
- The electrostatic potential varies in the opposite direction, and can be given by V(x) = $E_i$(x)/(-q).
- Thus, the electric field can be given by

$$E(x) = -\frac{dV(x)}{dx} = \frac{1}{q}\frac{dE_i(x)}{dx}$$

- Note: electrons drift downhill in a band diagram, therefore, the electric field points uphill in a band diagram.
- Note: in equilibrium, no net current => any fluctuation in carrier concentration which brings about a diffusion current also sets up an electric field, which opposes the diffusive motion => thus, equilibrium is established.
- This field is referred to as the built-in field, and can be caused by doping gradients and/or variation in the band gap.
- Equating the hole current density equation to zero, we get

$$E(x) = \frac{D_p}{\mu_p}\frac{1}{p(x)}\frac{dp(x)}{dx} = \frac{D_p}{\mu_p}\frac{1}{kT}\left(\frac{dE_i}{dx} - \frac{dE_F}{dx}\right)$$

- Now, EF does not vary with x in equilibrium, and the variation of $E_i$ with x is given above, thus,

$$\frac{D}{\mu} = \frac{kT}{q}$$

- This is an extremely important equation valid for both carrier types, and is called the Einstein relation gives the relation between D and , which is a function only of temperature, and allows calculation of one if the other is known.

## Diffusion and Recombination: The Continuity Equation

- In the description of conduction processes, the effects of recombination must be included, since they can cause a variation in the carrier distribution.
- Hole conservation equation:

$$\left.\frac{\partial p}{\partial t}\right|_{x \to x+\Delta x} = \frac{1}{q}\frac{J_p(x) - J_p(x+\Delta x)}{\Delta x} - \frac{\delta p}{\tau_p}$$

i.e., rate of hole buildup = increase of hole concentration in $\Delta xA$ per unit time recombination rate.

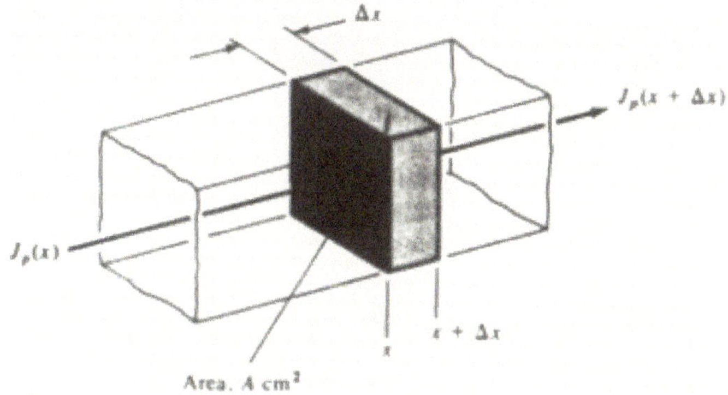


Setup to obtain particle count: current entering and leaving a volume $\Delta xA$.

- For $\Delta x \to 0$,

$$\frac{\partial p(x,t)}{\partial t} = \frac{\partial \delta p}{\partial t} = -\frac{1}{q}\frac{\partial J_p}{\partial x} - \frac{\delta p}{\tau_p}$$

- This is called the continuity equation for holes, and, similarly, for electrons, we can write

$$\frac{\partial \delta n}{\partial t} = \frac{1}{q}\frac{\partial J_n}{\partial x} - \frac{\delta n}{\tau_n}$$

- When the current is carried entirely due to diffusion (negligible drift), then we obtain the diffusion equation for electrons, given by $\frac{\partial \delta n}{\partial t} = D_n \frac{\partial^2 \delta n}{\partial x^2} - \frac{\delta n}{\tau_n}$

  and, similarly for holes,

$$\frac{\partial \delta p}{\partial t} = D_p \frac{\partial^2 \delta p}{\partial x^2} - \frac{\delta p}{\tau_p}$$

## Steady State Carrier Injection: Diffusion Length

- In steady state, if a distribution of excess carriers is maintained, then the diffusion equations become

$$\frac{\partial^2 \delta n}{\partial x^2} = \frac{\delta n}{L_n^2} \quad \textit{and} \quad \frac{\partial^2 \delta p}{\partial x^2} = \frac{\delta p}{L_p^2}$$

  where $L_n (\equiv \sqrt{D_n \tau_n})$ *and* $L_p (\equiv \sqrt{D_p \tau_p})$

  are the electron diffusion length and the hole diffusion length, which is the average distance an electron or a hole diffuses before recombining respectively.

- Case study: suppose excess holes are injected into a semi-infinite n-type bar, which maintains a constant concentration $\delta p(x=0) = \Delta p$ (relevant problem in a forward biased diode).

- Obviously, the excess holes would diffuse into the n-type bar, recombine with the electrons with a characteristic lifetime $\tau_p$ (and diffusion length $L_p$), and for large values of x, the excess hole concentration should decay to zero; thus,

  $$\delta p(x) = \Delta p e^{-x/L_p}$$

  and the decay profile is exponential.

- The steady state distribution of excess holes causes diffusion, and therefore, a hole current in the direction of decreasing concentration, given by

  $$J_p(x) = -qD_p \frac{dp}{dx} = -qD_p \frac{d\delta p}{dx} = q\frac{D_p}{L_p}\Delta p e^{-x/L_p} = q\frac{D_p}{L_p}\delta p(x)$$

  (This equation would come handy in the diode analysis.)

## The Haynes-Shockley Experiment

- Counterpart of the Hall effect experiment.
- Independently determines the minority carrier mobility and diffusion coefficient

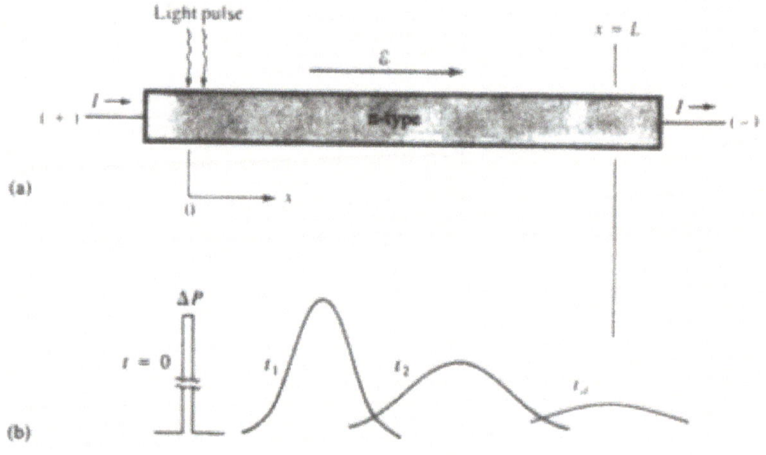


Schematic for Haynes-Shockley experiment: drift and diffusion of a hole pulse in an n-type bar:

(a) sample geometry,

(b) shape and position of the pulse for different times along the bar.

## Basic Principle

- a pulse of excess holes is created in an n-type bar (which has an applied electric field),
- as time progresses, the holes spread out by diffusion and move due to the electric field, and their motion is monitored somewhere down the bar,
- the time required for the holes to move the distance gives a measure of their mobility, and the spreading of the pulse in a given time gives a measure of the diffusion coefficient.

- A pulse of excess carriers is created by a light flash at x = 0 in an n-type semiconductor bar with an electric field E.
- The excess holes drift down the bar and reach the point x = L after a time , thus, the drift velocity $V_d = L / t_d$, and the hole mobility $\mu_p = V_d / E$.
- For measurement of diffusion coefficient, assume the pulse spreads without drift and neglect recombination; then the diffusion equation can be rewritten as

$$\frac{\partial \delta p(x,t)}{\partial t} = D_p \frac{\partial^2 \delta p(x,t)}{\partial x^2}$$

- The solution to this equation is the well known Gaussian distribution, given by

$$\delta p(x,t) = \left[\frac{\Delta p}{2\sqrt{\pi D_p t}}\right] e^{-x^2/4D_p t}$$

where the number of holes per unit area created over a negligibly small distance at t = 0.

- Note: the peak values of the pulse decreases and the pulse spreads in $\pm x$ directions with time.
- Let the peak value of the pulse be at $\hat{\delta} p \ at \ t = t_d$ and note that at $x = \Delta x / 2$, $\delta p$ is down by 1/e of its peak value $\hat{\delta} p$; thus,

$$e^{-1} \delta \hat{p} = \hat{\delta} p e^{-(\Delta x/2)^2 / 4D_p t_d}$$

and $D_p = (\Delta x)^2 / 16 t_d$, where $\Delta x = \Delta t . v_d = \Delta t . L / t_d$, where t is the spread of the pulse seen in an oscilloscope in time.

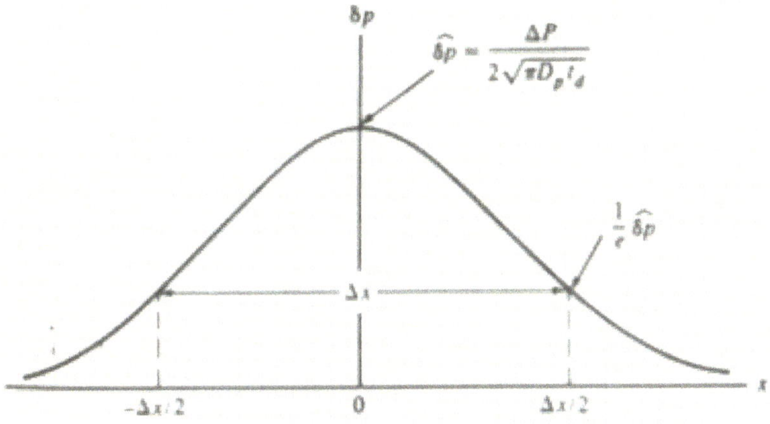

The profile of the excess hole concentration after time td.

## Gradients in the Imrefs

- Any combination of drift and diffusion implies a gradient in the steady state Imrefs.
- Under general case of nonequilibrium electron concentration with drift and diffusion, the total electron current can be written as

$$J_n(x) = q\mu_n n(x)E(x) + qD_n \frac{dn(x)}{dx}$$

- Using the expression for n(x) in terms of the electron Imref, and applying Einstein relation, it can be shown that

$$J_n(x) = \mu_n n(x)\frac{dF_n}{dx}$$

and, similarly, for holes,

$$J_p(x) = \mu_p p(x)\frac{dF_p}{dx}$$

Therefore, any drift, diffusion, or a combination of the two in a semiconductor sets up currents proportional to the gradient in the Imrefs, or, in other words, no current implies constant Imrefs.

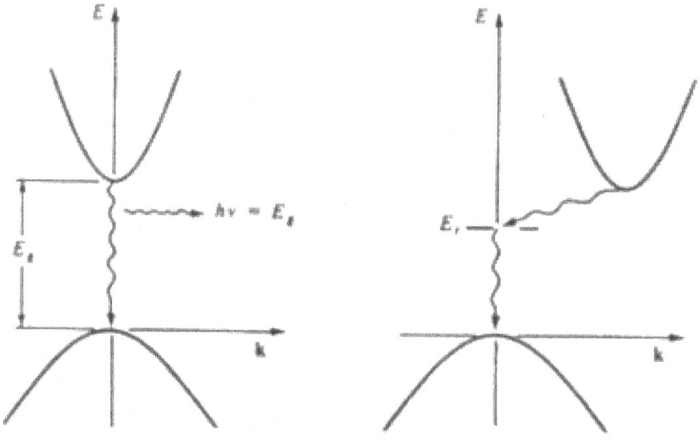


Fluorescent solutions under UV-light. Absorbed photons are rapidly re-emitted under longer wavelength.

Photoluminescence (abbreviated as PL) is light emission from any form of matter after the absorption of photons (electromagnetic radiation). It is one of many forms of luminescence (light emission) and is initiated by photoexcitation (excitation by photons), hence the prefix *photo-*. Following excitation various relaxation processes typically occur in which other photons are re-radiated. Time periods between absorption and emission may vary: ranging from short femtosecond-regime for emission involving free-carrier plasma in inorganic semiconductors up to milliseconds for phosphorescent processes in molecular systems; and under special circumstances delay of emission may even span to minutes or hours.

Observation of photoluminescence at a certain energy can be viewed as indication that excitation populated an excited state associated with this transition energy.

While this is generally true in atoms and similar systems, correlations and other more complex phenomena also act as sources for photoluminescence in many-body systems such as semiconductors. A theoretical approach to handle this is given by the semiconductor luminescence equations.

## Forms of Photoluminescence

Photoluminescence processes can be classified by various parameters such as the energy of the exciting photon with respect to the emission. Resonant excitation describes a situation in which photons of a particular wavelength are absorbed and equivalent photons are very rapidly re-emitted. This is often referred to as resonance fluorescence. For materials in solution or in the gas phase, this process involves electrons but no significant internal energy transitions involving molecular features of the chemical substance between absorption and emission. In crystalline inorganic semiconductors where an electronic band structure is formed, secondary emission can be more complicated as events may contain both coherent contributions such as resonant Rayleigh scattering where a fixed phase relation with the driving light field is maintained (i.e. energetically elastic processes where no losses are involved), and incoherent contributions (or inelastic modes where some energy channels into an auxiliary loss mode),

The latter originate, e.g., from the radiative recombination of excitons, Coulomb-bound electron-hole pair states in solids. Resonance fluorescence may also show significant quantum optical correlations.

More processes may occur when a substance undergoes internal energy transitions before re-emitting the energy from the absorption event. Electrons change energy states by either resonantly gaining energy from absorption of a photon or losing energy by emitting photons. In chemistry-related disciplines, one often distinguishes between fluorescence and phosphorescence. The prior is typically a fast process, yet some amount of the original energy is dissipated so that re-emitted light photons will have lower energy than did the absorbed excitation photons. The re-emitted photon in this case is said to be red shifted, referring to the reduced energy it carries following this loss (as the Jablonski diagram shows). For phosphorescence, absorbed photons undergo intersystem crossing where they enter into a state with altered spin multiplicity, usually a triplet state. Once energy from this absorbed electron is transferred in this triplet state, electron transition back to the lower singlet energy states is quantum mechanically forbidden, meaning that it happens much more slowly than other transitions. The result is a slow process of radiative transition back to the singlet state, sometimes lasting minutes or hours. This is the basis for "glow in the dark" substances.

Photoluminescence is an important technique for measuring the purity and crystalline quality of semiconductors such as GaAs and InP and for quantification of the amount of disorder present in a system. Several variations of photoluminescence exist, including photoluminescence excitation (PLE) spectroscopy.

Time-resolved photoluminescence (TRPL) is a method where the sample is excited with a light pulse and then the decay in photoluminescence with respect to time is measured. This technique is useful for measuring the minority carrier lifetime of III-V semiconductors like gallium arsenide (GaAs).

## Photoluminescence Properties of Direct-gap Semiconductors

In a typical PL experiment, a semiconductor is excited with a light-source that provides photons with an energy larger than the bandgap energy. The incoming light excites a polarization that can be described with the semiconductor Bloch equations. Once the photons are absorbed, electrons and holes are formed with finite momenta k in the conduction and valence bands, respectively. The excitations then undergo energy and momentum relaxation towards the band gap minimum. Typical mechanisms are Coulomb scattering and the interaction with phonons. Finally, the electrons recombine with holes under emission of photons.

Ideal, defect-free semiconductors are many-body systems where the interactions of charge-carriers and lattice vibrations have to be considered in addition to the light-matter coupling. In general, the PL properties are also extremely sensitive to internal electric fields and to the dielectric environment (such as in photonic crystals) which impose further degrees of complexity. A precise microscopic description is provided by the semiconductor luminescence equations.

## Ideal Quantum-well Structures

An ideal, defect-free semiconductor quantum well structure is a useful model system to illustrate the fundamental processes in typical PL experiments. The discussion is based on results published in Klingshirn (2012) and Balkan (1998).

The fictive model structure for this discussion has two confined quantized electronic and two hole subbands, $e_1$, $e_2$ and $h_1$, $h_2$, respectively. The linear absorption spectrum of such a structure shows the exciton resonances of the first (e1h1) and the second quantum well subbands ($e_2$, $h_2$), as well as the absorption from the corresponding continuum states and from the barrier.

## Photoexcitation

In general, three different excitation conditions are distinguished: resonant, quasi-resonant, and non-resonant. For the resonant excitation, the central energy of the laser corresponds to the lowest exciton resonance of the quantum well. No or only a negligible amount of the excess energy is injected to the carrier system. For these conditions, coherent processes contribute significantly to the spontaneous emission. The decay of polarization creates excitons directly. The detection of PL is challenging for resonant excitation as it is difficult to discriminate contributions from the excitation, i.e., stray-light and diffuse scattering from surface roughness. Thus, speckle and resonant Rayleigh-scattering are always superimposed to the incoherent emission.

In case of the non-resonant excitation, the structure is excited with some excess energy. This is the typical situation used in most PL experiments as the excitation energy can be discriminated using a spectrometer or an optical filter. One has to distinguish between quasi-resonant excitation and barrier excitation.

For quasi-resonant conditions, the energy of the excitation is tuned above the ground state but still below the barrier absorption edge, for example, into the continuum of the first subband. The polarization decay for these conditions is much faster than for resonant excitation and coherent contributions to the quantum well emission are negligible. The initial temperature of the carrier system is significantly higher than the lattice temperature due to the surplus energy of the injected carriers. Finally, only the electron-hole plasma is initially created. It is then followed by the formation of excitons.

In case of barrier excitation, the initial carrier distribution in the quantum well strongly depends on the carrier scattering between barrier and the well.

## Relaxation

Initially, the laser light induces coherent polarization in the sample, i.e., the transitions between electron and hole states oscillate with the laser frequency and a fixed phase. The polarization dephases typically on a sub-100 fs time-scale in case of nonresonant excitation due to ultra-fast Coulomb- and phonon-scattering.

The dephasing of the polarization leads to creation of populations of electrons and holes in the conduction and the valence bands, respectively. The lifetime of the carrier populations is rather long, limited by radiative and non-radiative recombination such as Auger recombination. During this lifetime a fraction of electrons and holes may form excitons, this topic is still controversially discussed in the literature. The formation rate depends on the experimental conditions such as lattice temperature, excitation density, as well as on the general material parameters, e.g., the strength of the Coulomb-interaction or the exciton binding energy.

The characteristic time-scales are in the range of hundreds of picoseconds in GaAs; they appear to be much shorter in wide-gap semiconductors.

Directly after the excitation with short (femtosecond) pulses and the quasi-instantaneous decay of the polarization, the carrier distribution is mainly determined by the spectral width of the excitation, e.g., a laser pulse. The distribution is thus highly non-thermal and resembles a Gaussian distribution, centered at a finite momentum. In the first hundreds of femtoseconds, the carriers are scattered by phonons, or at elevated carrier densities via Coulomb-interaction. The carrier system successively relaxes to the Fermi–Dirac distribution typically within the first picosecond. Finally, the carrier system cools down under the emission of phonons. This can take up to several nanoseconds, depending on the material system, the lattice temperature, and the excitation conditions such as the surplus energy.

Initially, the carrier temperature decreases fast via emission of optical phonons. This is quite efficient due to the comparatively large energy associated with optical phonons, (36meV or 420K in GaAs) and their rather flat dispersion, allowing for a wide range of scattering processes under conservation of energy and momentum. Once the carrier temperature decreases below the value corresponding to the optical phonon energy, acoustic phonons dominate the relaxation. Here, cooling is less efficient due their dispersion and small energies and the temperature decreases much slower beyond the first tens of picoseconds. At elevated excitation densities, the carrier cooling is further inhibited by the so-called hot-phonon effect. The relaxation of a large number of hot carriers leads to a high generation rate of optical phonons which exceeds the decay rate into acoustic phonons. This creates a non-equilibrium "over-population" of optical phonons and thus causes their increased reabsorption by the charge-carriers significantly suppressing any cooling. A system thus cools slower, the higher the carrier density is.

## Radiative Recombination

The emission directly after the excitation is spectrally very broad, yet still centered in the vicinity of the strongest exciton resonance. As the carrier distribution relaxes and cools, the width of the PL peak decreases and the emission energy shifts to match the ground state of the exciton for ideal samples without disorder. The PL spectrum approaches its quasi-steady-state shape defined by the distribution of electrons and holes. Increasing the excitation density will change the emission spectra. They are dominated by the excitonic ground state for low densities. Additional peaks from higher subband transitions appear as the carrier density or lattice temperature are increased as these states get more and more populated. Also, the width of the main PL peak increases significantly with rising excitation due to excitation-induced dephasing and the emission peak experiences a small shift in energy due to the Coulomb-renormalization and phase-filling.

In general, both exciton populations and plasma, uncorrelated electrons and holes, can act as sources for photoluminescence as described in the semiconductor-luminescence equations. Both yield very similar spectral features which are difficult to distinguish; their emission dynamics, however, vary significantly. The decay of excitons yields a single-exponential decay function since the probability of their radiative recombination does not depend on the carrier density. The probability of spontaneous emission for uncorrelated electrons and holes, is approximately proportional to the product of electron and hole populations eventually leading to a non-single-exponential decay described by a hyperbolic function.

### Effects of Disorder

Real material systems always incorporate disorder. Examples are structural defects in the lattice or disorder due to variations of the chemical composition. Their treatment is extremely challenging for microscopic theories due to the lack of detailed knowledge about perturbations of the ideal structure. Thus, the influence of the extrinsic effects on the PL is usually addressed phenomenologically. In experiments, disorder can lead to localization of carriers and hence drastically increase the photoluminescence life times as localized carriers cannot as easily find nonradiative recombination centers as can free ones.

### Photoluminescent Material for Temperature Detection

In phosphor thermometry, the temperature dependence of the photoluminescence process is exploited to measure temperature.

### Experimental Methods

Photoluminescence spectroscopy is a widely used technique for characterisation of the optical and electronic properties of semiconductors and molecules. In chemistry, it is more often referred to as fluorescence spectroscopy, but the instrumentation is the same. The relaxation processes can be studied using Time-resolved fluorescence spectroscopy to find the decay lifetime of the photoluminescence. These techniques can be combined with microscopy, to map the intensity (Confocal microscopy) or the lifetime (Fluorescence-lifetime imaging microscopy) of the photoluminescence across a sample (e.g. a semiconducting wafer, or a biological sample that has been marked with fluorescent molecules).

## Energy Band Theory

Semiconductors Si, Ge, GaAs, InP, CdTe.. - all are crystalline solids.

This is a 3D regular arrangement of atoms called the lattice. Some elemental semiconductors such as Si & Ge have only one kind of material at the lattice sites, where as GaAs, InP,.are compound semiconductors and are composed of two interpenetrating 3D lattices.

There may be defects in the lattice and each atom in a lattice vibrates.

Defects and vibrations are second order phenomenon which would be discussed separately.

If we have a charged ion at x = 0 and an electron outside the attractive force between the ion and the electron is

$$F = \frac{Z_q^2}{4\pi\varepsilon_0 r^2}$$

The PE of the $ion = -\frac{Z_q^2}{4\pi\varepsilon_0 r}$ $or$ $-\frac{Z_q^2}{4\pi\varepsilon_0 x}$ and PE =0 at $x = \pm\infty$

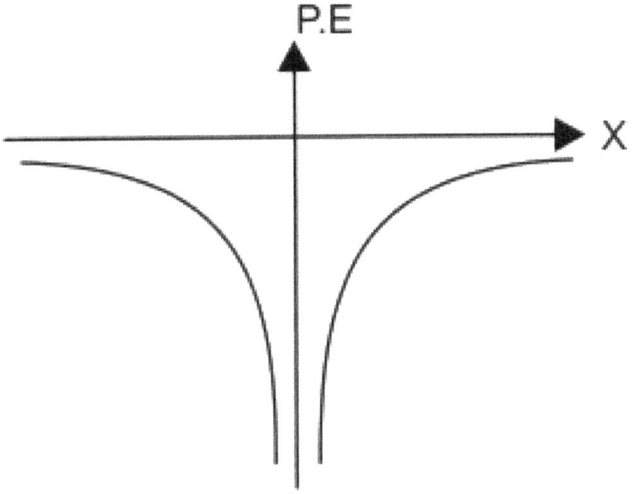


If we consider two ions separated by 'a', the potential profile in on shown in Figure.

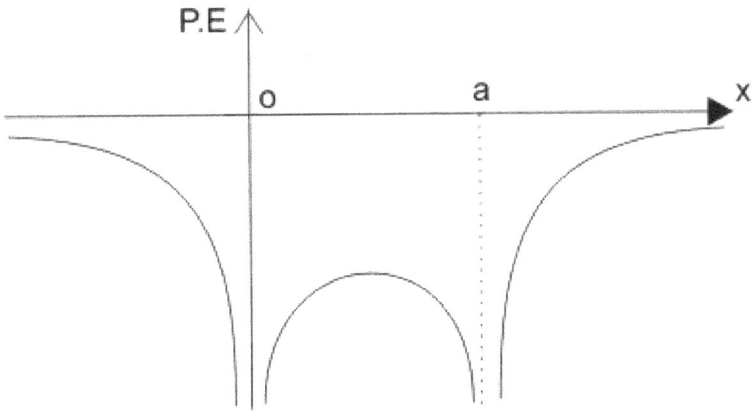


To get a feel for the electron energies in a crystal semiconductor let us consider an electron in a 1D lattice of atoms. The simplest lattice structure is a 1 D lattice.

A regular array of atoms placed periodically.

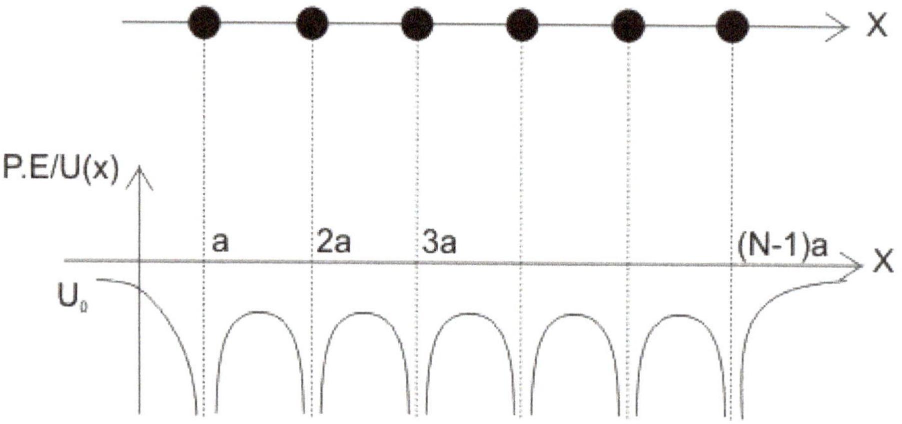


At each lattice site (x = 0, a, 2a, ..) there exists an ion with some net charge $Z_q$

## References


- Justin Iveland; Lucio Martinelli; Jacques Peretti; James S. Speck; Claude Weisbuch. "Cause of LED Efficiency Droop Finally Revealed". Physical Review Letters, 2013. Science Daily. Retrieved 23 April 2013
- Shockley, W.; Read, W. T. (1 September 1952). "Statistics of the Recombinations of Holes and Electrons". Physical Review. 87 (5): 835–842. Bibcode:1952PhRv...87..835S. doi:10.1103/PhysRev.87.835
- Balkan, Naci (1998). Hot Electrons in Semiconductors: Physics and Devices. Oxford University Press. ISBN 0198500580
- Hayes, G.R.; Deveaud, B. (2002). "Is Luminescence from Quantum Wells Due to Excitons?". physica status solidi (a) 190 (3): 637–640. doi:10.1002/1521-396X(200204)190:3<637::AID-PSSA637>3.0.CO;2-7
- Kimble, H. J.; Dagenais, M.; Mandel, L. (1977). "Photon Antibunching in Resonance Fluorescence". Physical Review Letters 39 (11): 691–695. doi:10.1103/PhysRevLett.39.691

4

# Understanding P-N Junction

Semiconductors can be of two types, p type and n type. The boundary found inside a single crystal of a semiconductor which binds them together is known as a p-n junction. This chapter discusses P-N junction in a critical manner providing key analysis to the subject matter.

## p–n junction

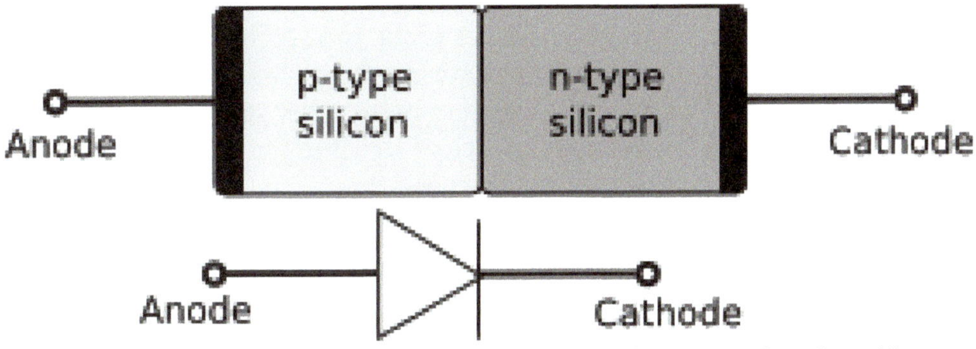


A p–n junction. The circuit symbol is shown: the triangle corresponds to the p side.

A p–n junction is a boundary or interface between two types of semiconductor material, p-type and n-type, inside a single crystal of semiconductor. The "p" (positive) side contains an excess of holes, while the "n" (negative) side contains an excess of electrons. The p-n junction is created by doping, for example by ion implantation, diffusion of dopants, or by epitaxy (growing a layer of crystal doped with one type of dopant on top of a layer of crystal doped with another type of dopant). If two separate pieces of material were used, this would introduce a grain boundary between the semiconductors that would severely inhibit its utility by scattering the electrons and holes.

p–n junctions are elementary "building blocks" of most semiconductor electronic devices such as diodes, transistors, solar cells, LEDs, and integrated circuits; they are the active sites where the electronic action of the device takes place. For example, a common type of transistor, the bipolar junction transistor, consists of two p–n junctions in series, in the form n–p–n or p–n–p.

The invention of the p–n junction is usually attributed to American physicist Russell Ohl of Bell Laboratories. However, Vadim Lashkaryov reported discovery of *p-n*-junctions in $Cu_2O$ and silver sulphide photocells and selenium rectifiers in 1941.

A Schottky junction is a special case of a p–n junction, where metal serves the role of the p-type semiconductor.

## Properties

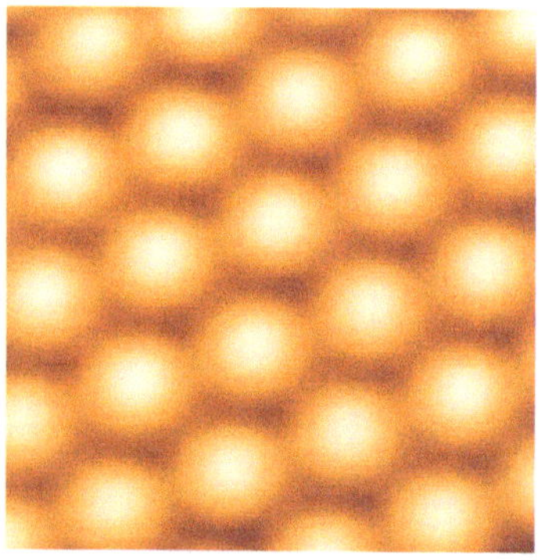

Image silicon atoms (Si) enlarged about 45,000,000x.

The p–n junction possesses some interesting properties that have useful applications in modern electronics. A p-doped semiconductor is relatively conductive. The same is true of an n-doped semiconductor, but the junction between them can become depleted of charge carriers, and hence non-conductive, depending on the relative voltages of the two semiconductor regions. By manipulating this non-conductive layer, p–n junctions are commonly used as diodes: circuit elements that allow a flow of electricity in one direction but not in the other (opposite) direction. *Bias* is the application of a voltage across a p–n junction; *forward bias* is in the direction of easy current flow, and *reverse bias* is in the direction of little or no current flow.

## Equilibrium (Zero Bias)

In a p–n junction, without an external applied voltage, an equilibrium condition is reached in which a potential difference is formed across the junction. This potential difference is called built-in potential $V_{bi}$.

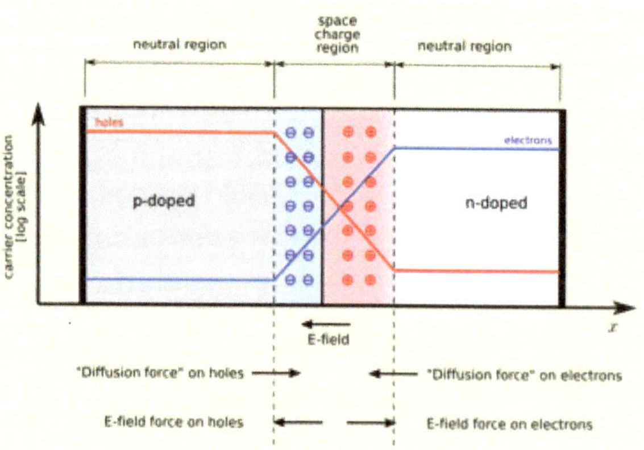


Figure A. A p–n junction in thermal equilibrium with zero-bias voltage applied. Electron and hole concentration are reported with blue and red lines, respectively. Gray regions are charge-neutral. Light-red zone is positively charged. Light-blue zone is negatively charged. The electric field is shown on the bottom, the electrostatic force on electrons and holes and the direction in which the diffusion tends to move electrons and holes. (The log concentration curves should actually be smoother with slope varying with field strength.)

After joining p-type and n-type semiconductors, electrons from the n region near the p–n interface tend to diffuse into the p region leaving behind positively charged ions in the n region and being recombined with holes, forming negatively charged ions in the p region. Likewise, holes from the p-type region near the p–n interface begin to diffuse into the n-type region, leaving behind negatively charged ions in the p region and recombining with electrons, forming positive ions in the n region Template:Explain holes move?. The regions near the p–n interface lose their neutrality and most of their mobile carriers, forming the space charge region or depletion layer.

The electric field created by the space charge region opposes the diffusion process for both electrons and holes. There are two concurrent phenomena: the diffusion process that tends to generate more space charge, and the electric field generated by the space charge that tends to counteract the diffusion. The carrier concentration profile at equilibrium is shown in figure A with blue and red lincs. Also shown are the two counterbalancing phenomena that establish equilibrium.

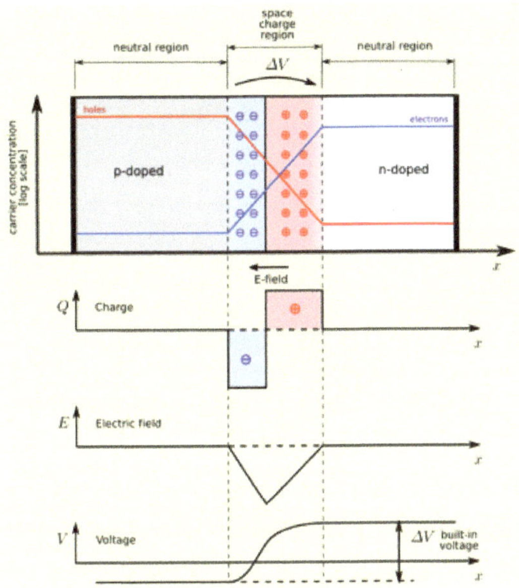


Figure B. A p–n junction in thermal equilibrium with zero-bias voltage applied. Under the junction, plots for the charge density, the electric field, and the voltage are reported. (The log concentration curves should actually be smoother, like the voltage.)

The space charge region is a zone with a net charge provided by the fixed ions (donors or acceptors) that have been left *uncovered* by majority carrier diffusion. When equilibrium is reached, the charge density is approximated by the displayed step function. In fact, since the y-axis of figure A is log-scale, the region is almost completely depleted of majority carriers (leaving a charge density equal to the net doping level), and the edge between the space charge region and the neutral region is quite sharp. The space charge region has the same magnitude of charge on both sides of the p–n interfaces, thus it extends farther on the less doped side in this example (the n side in figures A and B).

## Forward Bias

In forward bias, the p-type is connected with the positive terminal and the n-type is connected with the negative terminal.

With a battery connected this way, the holes in the p-type region and the electrons in the n-type region are pushed toward the junction and start to neutralize the depletion zone, reducing its width. The positive potential applied to the p-type material repels the holes, while the negative potential applied to the n-type material repels the electrons. The change in potential between the p side and the n side decreases or switches sign. With increasing forward-bias voltage, the depletion zone eventually becomes thin enough that the zone's electric field cannot counteract charge carrier motion across the p–n junction, which as a consequence reduces electrical resistance. The electrons that cross the p–n junction into the p-type material (or holes that cross into the n-type material) will diffuse into the nearby neutral region. The amount of minority diffusion in the near-neutral zones determines the amount of current that may flow through the diode.

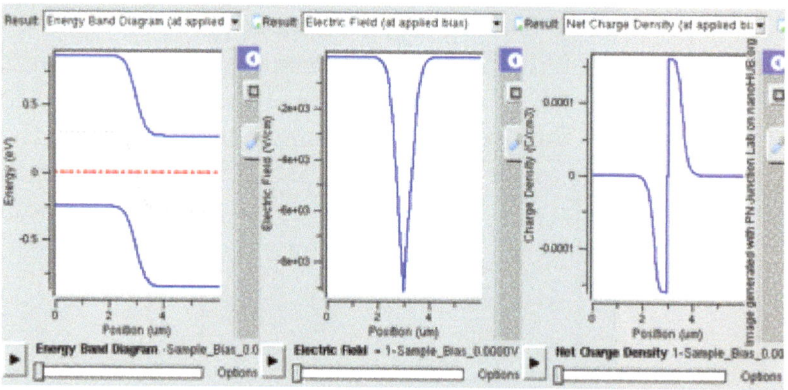


PN junction operation in forward-bias mode, showing reducing depletion width. The panels show energy band diagram, electric field, and net charge density. Both p and n junctions are doped at a 1e15/cm3 (0.00016C/cm³) doping level, leading to built-in potential of ~0.59 V. Reducing depletion width can be inferred from the shrinking charge profile, as fewer dopants are exposed with increasing forward bias. Observe the different quasi-fermi levels for conduction band and valence band in n and p regions (red curves)

Only majority carriers (electrons in n-type material or holes in p-type) can flow through a semiconductor for a macroscopic length. With this in mind, consider the flow of electrons across the junction. The forward bias causes a force on the electrons pushing them from the N side toward the P side. With forward bias, the depletion region is narrow enough that electrons can cross the junction and *inject* into the p-type material. However, they do not continue to flow through the p-type material indefinitely, because it is energetically favorable for them to recombine with holes. The average length an electron travels through the p-type material before recombining is called the *diffusion length*, and it is typically on the order of micrometers.

Although the electrons penetrate only a short distance into the p-type material, the electric current continues uninterrupted, because holes (the majority carriers) begin to flow in the opposite direction. The total current (the sum of the electron and hole currents) is constant in space, because any variation would cause charge buildup over time (this is Kirchhoff's current law). The flow of holes from the p-type region into the n-type region is exactly analogous to the flow of electrons from N to P (electrons and holes swap roles and the signs of all currents and voltages are reversed).

Therefore, the macroscopic picture of the current flow through the diode involves electrons flowing through the n-type region toward the junction, holes flowing through the p-type region in the opposite direction toward the junction, and the two species of carriers constantly recombining in the vicinity of the junction. The electrons and holes travel in opposite directions, but they also have opposite charges, so the overall current is in the same direction on both sides of the diode, as required.

The Shockley diode equation models the forward-bias operational characteristics of a p–n junction outside the avalanche (reverse-biased conducting) region.

## Reverse Bias

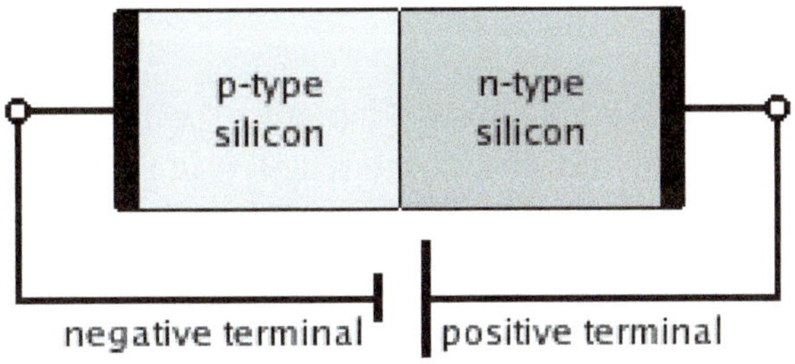


A silicon p–n junction in reverse bias.

Connecting the *p-type* region to the *negative* terminal of the battery and the *n-type* region to the *positive* terminal corresponds to reverse bias. If a diode is reverse-biased, the voltage at the cathode is comparatively higher than at the anode. Therefore, very little current will flow until the diode breaks down. The connections are illustrated in the adjacent diagram.

Because the p-type material is now connected to the negative terminal of the power supply, the 'holes' in the p-type material are pulled away from the junction, leaving behind charged ions and causing the width of the depletion region to increase. Likewise, because the n-type region is connected to the positive terminal, the electrons will also be pulled away from the junction, with similar effect. This increases the voltage barrier causing a high resistance to the flow of charge carriers, thus allowing minimal electric current to cross the p–n junction. The increase in resistance of the p–n junction results in the junction behaving as an insulator.

The strength of the depletion zone electric field increases as the reverse-bias voltage increases. Once the electric field intensity increases beyond a critical level, the p–n junction depletion zone breaks down and current begins to flow, usually by either the Zener or the avalanche breakdown processes. Both of these breakdown processes are non-destructive and are reversible, as long as the amount of current flowing does not reach levels that cause the semiconductor material to overheat and cause thermal damage.

This effect is used to advantage in Zener diode regulator circuits. Zener diodes have a low breakdown voltage. A standard value for breakdown voltage is for instance 5.6 V. This means that the voltage at the cathode cannot be more than about 5.6 V higher than the voltage at the anode (although there is a slight rise with current), because the diode will break down – and therefore conduct – if the voltage gets any higher. This in effect limits the voltage over the diode.

Another application of reverse biasing is Varicap diodes, where the width of the depletion zone (controlled with the reverse bias voltage) changes the capacitance of the diode.

## Governing Equations

### Size of Depletion Region

For a p–n junction, letting $C_A(x)$ and $C_D(x)$ be the concentrations of acceptor and donor atoms respectively, and letting $N_0(x)$ and $P_0(x)$ be the equilibrium concentrations of electrons and holes respectively, yields, by Poisson's equation:

$$-\frac{d^2V}{dx^2}=\frac{\rho}{\varepsilon}=\frac{q}{\varepsilon}[(N_0-P_0)+(C_D-C_A)]$$

where $V$ is the electric potential, $\rho$ is the charge density, $\varepsilon$ is permittivity and $q$ is the magnitude of the electron charge. Letting $d_p$ be the width of the depletion region within the p-side, and letting $d_n$ be the width of the depletion region within the n-side, it must be that

$$d_pC_A=d_nC_D$$

because the total charge on either side of the depletion region must cancel out. Therefore, letting $D$ and $\Delta V$ represent the entire depletion region and the potential difference across it,

$$\Delta V=\int_D\int\frac{q}{\varepsilon}[(N_0-P_0)+(C_D-C_A)]dxdx$$

$$=\frac{C_AC_D}{C_A+C_D}\frac{q}{2\varepsilon}(d_p+d_n)^2$$

where $P_0=N_0=0$, because we are in the depletion region. And thus, letting $d$ be the total width of the depletion region, we get

$$d=\sqrt{\frac{2\varepsilon}{q}\frac{C_A+C_D}{C_AC_D}\Delta V}$$

$\Delta V$ can be written as $\Delta V_0+\Delta V_{\text{ext}}$, where we have broken up the voltage difference into the equilibrium plus external components. The equilibrium potential results from diffusion forces, and thus we can calculate $\Delta V_0$ by implementing the Einstein relation and assuming the semiconductor is nondegenerate (i.e. the product $P_0N_0$ is independent of the Fermi energy):

$$\Delta V_0=\frac{kT}{q}\ln\left(\frac{C_AC_D}{P_0N_0}\right)$$

where $T$ is the temperature of the semiconductor and $k$ is Boltzmann constant.

## Current Across Depletion Region

The *Shockley ideal diode equation* characterizes the current across a p–n junction as a function of external voltage and ambient conditions (temperature, choice of semiconductor, etc.). To see how it can be derived, we must examine the various reasons for current. The convention is that the forward (+) direction be pointed against the diode's built-in potential gradient at equilibrium.

- Forward Current ($\mathbf{J}_F$)
  - Diffusion Current: current due to local imbalances in carrier concentration $n$, via the equation $\mathbf{J}_D\propto -q\nabla n$
- Reverse Current ($\mathbf{J}_R$)

- Field Current
- Generation Current

## Summary

The forward-bias and the reverse-bias properties of the p–n junction imply that it can be used as a diode. A p–n junction diode allows electric charges to flow in one direction, but not in the opposite direction; negative charges (electrons) can easily flow through the junction from n to p but not from p to n, and the reverse is true for holes. When the p–n junction is forward-biased, electric charge flows freely due to reduced resistance of the p–n junction. When the p–n junction is reverse-biased, however, the junction barrier (and therefore resistance) becomes greater and charge flow is minimal.

## Non-rectifying Junctions

In the above diagrams, contact between the metal wires and the semiconductor material also creates metal–semiconductor junctions called Schottky diodes. In a simplified ideal situation a semiconductor diode would never function, since it would be composed of several diodes connected back-to-front in series. But, in practice, surface impurities within the part of the semiconductor that touches the metal terminals will greatly reduce the width of those depletion layers to such an extent that the metal-semiconductor junctions do not act as diodes. These *non-rectifying junctions* behave as ohmic contacts regardless of applied voltage polarity.

## PN Junction in Equilibrium

- PN junctions are important for the following reasons:
  (i) PN junction is an important semiconductor device in itself and used in a wide variety of applications such as rectifiers, Photodetectors, light emitting diodes and lasers etc
  (ii) PN junctions are an integral part of other important semiconductor devices such as BJTs, JFETS and MOSFETs
  (iii) PN junctions are used as test structures for measuring important semiconductor properties such as doping, defect density, lifetime etc
- The discussion associated with the PN junctions will proceed in the following order:
  (i) PN junction in equilibrium
  (ii) dc IV characteristics in forward bias
  (iii)characteristics in reverse bias
  (iv) dynamic characteristics
  (v) Circuit models
  (vi) Design perspective

Device Structure :

The Figure below shows a simplified structure of a PN junction:

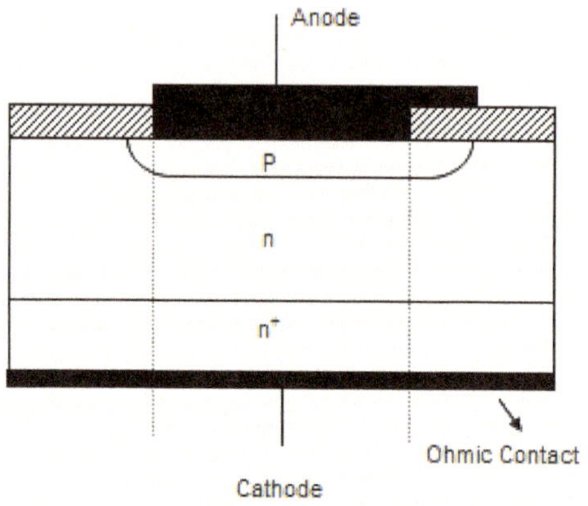


- The structure can be fabricated by diffusing P-type impurity in the n-epilayer grown over an $n^+$ substrate.
- While the doping in the n-epilayer can be uniform, the doping in the P-region is often either Gaussian or error-function in nature. The doping profiles and the junction are schematically illustrated below:

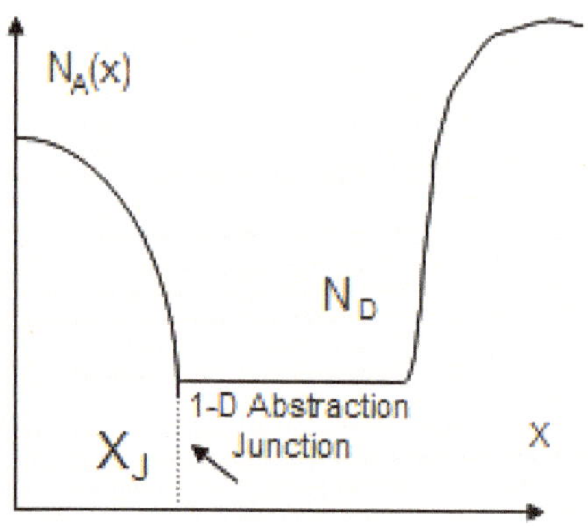


1-D Abstraction

- Even though the doping in both N and P-regions may in general be nonuniform, for simplicity, we shall assume them to be uniform in the initial analysis because the basic device physics remains almost the same
- A simplified, one-dimensional abstracted view of a PN junction described by the region within the dotted lines of device schematic is shown below:

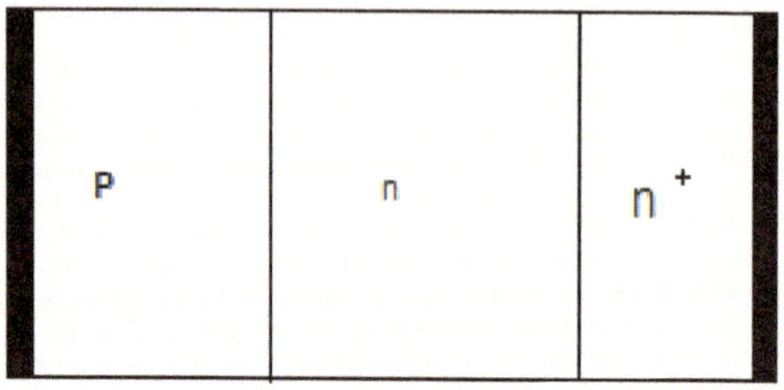

- We shall assume that the thicknesses of P and N-regions are large enough so that one can ignore the presence of Ohmic contacts and the heavily doped N-region and consider only the P and N regions for analysis. Such a diode with wide N and P-regions is called a wide-base diode.
- The PN junction that we shall study will therefore be a 1-D structure with uniformly doped P and N regions with thicknesses sufficiently large to ignore effects of contacts and other layers. It shall be represented simply as

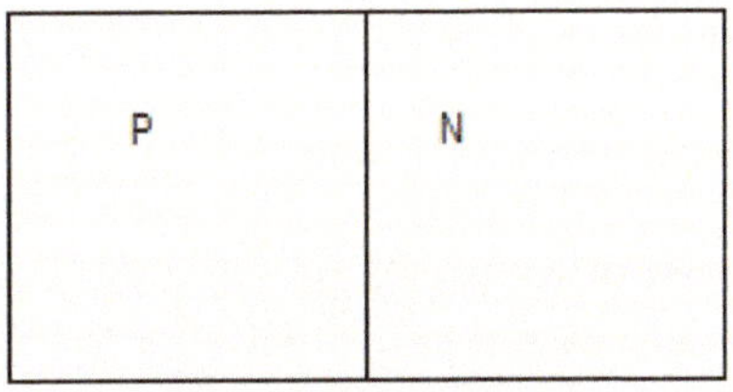


PN junction in Equilibrium

- As mentioned earlier, the characteristics of a semiconductor device is completely specified in equilibrium if the variation of potential as a function of position is specified. As a first step to obtaining this potential profile, we shall sketch the energy-band diagram of the device. The energy band diagram would provide us with

    (i) a qualitative variation of potential in the device

    (ii) boundary conditions for solution of Poisson's equation

- As usual, the energy band diagram of the PN junction will be obtained by combining the energy band diagrams of N and P-type semiconductors separately

## Energy Band Diagram In Equilibrium

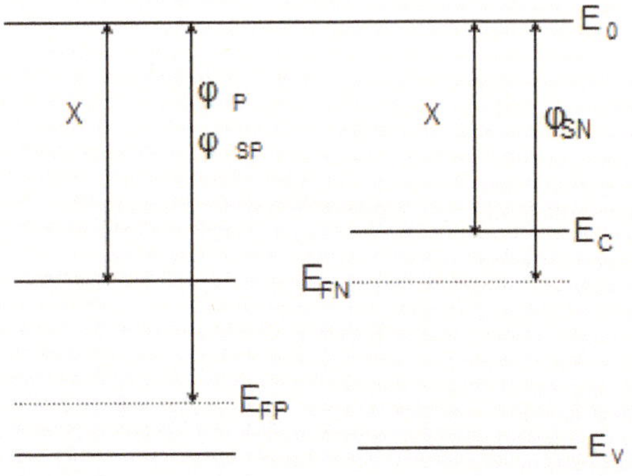


Energy Band diagram of N- and P-regions before equilibrium

- When the N and P-regions are brought into contact, the electrons would flow from regions of higher Fermi-energy to regions of lower Fermi energy and holes would flow in the opposite direction.
- Because of loss of electrons, the N-region would acquire a net positive charge due to the uncovered positively charged donor atoms and P-region would acquire a negative charge due to uncovered negatively charged acceptor atoms.

- At equilibrium there is no net flow of either electrons or holes so that the PN junction has a single constant Fermi level.
- The transfer of charges will affect only the regions close to the junction so that regions which are far still have the same energy band diagram(i.e. same relative positions of conduction and valence band wrt Fermi energy)
- As we approach the junction from the N-side, the conduction band must bend upwards away from the Fermi energy to indicate the fact that the region is progressively getting depleted of electrons ( remember $n_O = N_C \exp\left(-\frac{E_c - E_F}{kT}\right)$).
- Similarly, as we approach the junction from the P-side, the conduction band must bend downwards towards the Fermi energy to indicate the fact that the region is getting depleted of holes

Using these principles, the final energy band diagram can be sketched a

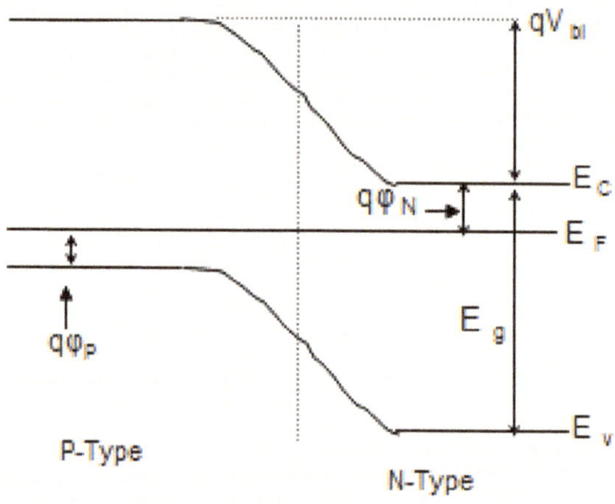


- As a result of transfer of charges from N and P-regions, the region next to the junction is charged and is known as the space charge region.
- The charge on the N-side is positive and on the P-side negative.
- As a result, the space charge region will have an electric field directed from the N to the P-region with a maximum value at the junction and zero at the edges of the space charge region.
- As a result of the electric field, there will be a net voltage across the space charge region known as the built-in voltage.
- The magnitude of the built-in voltage can be quickly estimated from the energy band diagram. We do this by performing an analog of Kirchoffs voltage law:

  We start from a point in the N-region(away from the space charge region) at the energy $E_F$ and then move to a point in the P-region(away from the space charge region)again at energy $E_F$ via any path other than the Fermi-energy and add up the energy gained or loss at each step of the path, then the net sum should be zero!

$$q\phi_N + qV_{bi} - E_g + q\phi_P = 0 \qquad (1)$$

The built-in oltage can be expessed as:

$$V_{bi} = \frac{E_g}{q} - \phi_N - \phi_P \qquad (2)$$

For Non-degenerate semiconductors:

$$\phi_N = \frac{kT}{q} In(\frac{N_C}{N_D})$$

$$\phi_P = \frac{kT}{q} In(\frac{N_V}{N_A}) \qquad (3)$$

- An important result that can be deduced from Eq.(2) is that built-in voltage will be higher for semiconductors with larger bandgap.

  Using the relationship $n_i^2 = N_C N_V \exp(-\frac{E_g}{kT})$, the expression for built-in voltage for a PN junction having non-degenerate semiconductors can be written as

$$V_{bi} = \frac{kT}{q} In(\frac{N_D N_A}{n_i^2}) \qquad (4)$$

- There is another method by which the magnitude of built-in voltage can be obtained. In this case we start with the fact that in equilibrium, the net electron current is zero

$$J_N = qn\mu_N \varepsilon_X + qD_N \frac{dn}{dx} = 0 \qquad (5)$$

Use of Einstein's relation : $D_N / \mu_N = kT / q$ allows the above expression to be re-written as:

$$\varepsilon_x = \frac{d\psi}{dx} = \frac{kT}{q} \frac{dln(x)}{dx} \qquad (6)$$

Integrating the above expression across the space charge region gives:

$$\int_{\psi_P}^{\psi} d\psi = \frac{kT}{q} \int_{n_P}^{n} \frac{dl\,n(x)}{dx} \qquad (7)$$

where $\psi_N$ $\psi_P$ are the potentials in the bulk of N- and P- regions respectively. $n_N = N_D$ is the electron density in the N-region and $n_P = \frac{n_i^2}{N_A}$ is the elecron density in the P-region.

$$\Rightarrow V_{bi} = \psi_N \qquad \psi_P = \frac{kT}{q} In(\frac{N_D N_A}{n_i^2}) \qquad (8)$$

## Poisson's Equation

- The energy band diagram gives only a qualitative variation of potential across the space charge region. The detailed nature of this potential can be obtained through the solution of Poisson equation:

$$\frac{\partial^2 \psi}{\partial x^2} = \frac{\rho}{\varepsilon_{Si}} \qquad (9)$$

$$\rho(x) = q(N_D^+ \quad N_A + p_O \quad n_O) \qquad (10)$$

$$n_0 = n_i \exp(\frac{q\psi}{kT}) \; ; \; p_0 = n_i \exp(\frac{q\psi}{kT}) \qquad (11)$$

## Analytical Solution of Poisson's Equation

- Because of the exponential terms in the expression for charge density, the analytical solution of the Poisson's equation becomes difficult.
- This difficulty is overcome through the assumption that the electron and hole density within the space charge region is negligible as compared to the ionized donor or acceptor atom density. This approximation, known as the depletion approximation, allows the Poisson equation to be simplified to:

$$\frac{\partial^2 \psi}{\partial x^2} = \frac{q}{\varepsilon_{Si}}(N_D^+ \quad N_A) \qquad (12)$$

Henceforth, we shall also assume that all donor and acceptor atoms are ionized.

- The table below shows the charge density as a function of potential within the space charge region for a PN junction with same doping ($10^{17} cm^{-3}$) in N and P regions for simplicity.

$$V_{bi} = \psi_N - \psi_P = 0.834$$

| $\psi - \psi_P$ | $\rho/qN_D$ |
|---|---|
| .01 | 0.32 |
| .02 | 0.54 |
| .06 | 0.9 |
| .1 | 0.98 |
| .5 | 1.0 |
| 0.7 | 1.0 |
| 0.774 | 0.9 |
| 0.814 | 0.54 |
| 0.824 | 0.32 |

- The data in the table shows that over a large range of potential, the depletion approximation is valid. Only for regions close to the space charge edge, does the approximation become weak

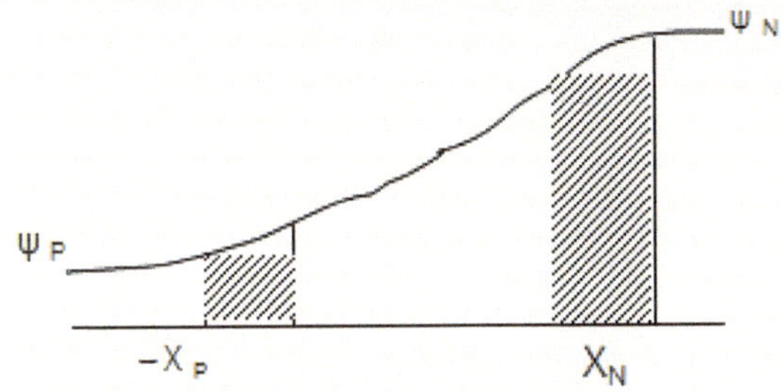


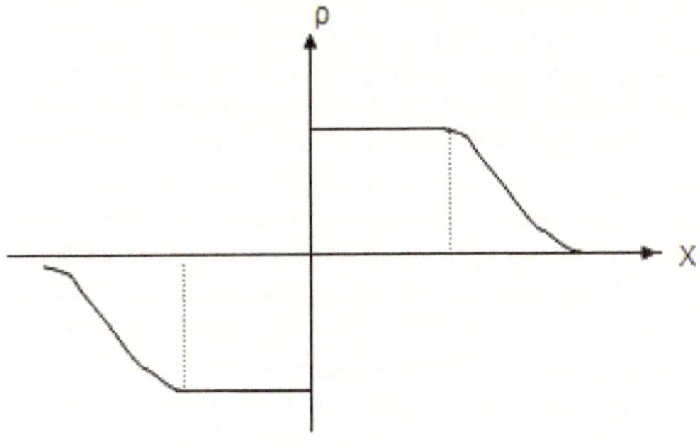


Simplified Charge density

With the depletion approximation , the charge density can be expressed as

$$p = qN_D \; for \; \text{Region} - \text{II}: \quad 0 \le x \le x_N$$

(13)

$$p = qN_A \; for \; \text{Region} - \text{I}: \quad -x_P \le x \le 0$$

The space charge region is often called the depletion region

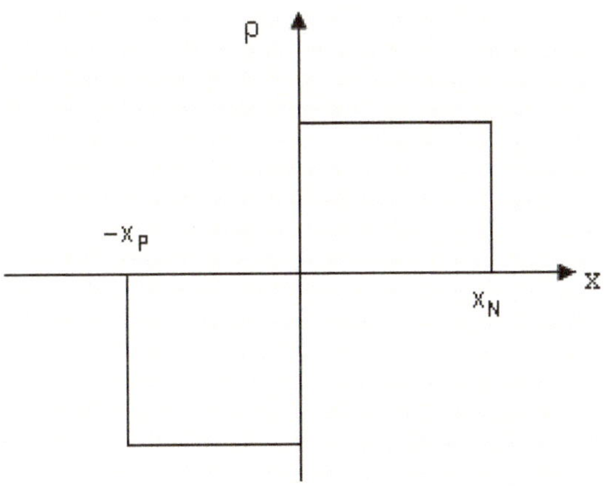


## Simplified Poisson Equation

The Poisson's equation for P and N-regions of the depletion region can be written a

$$\frac{\partial^2 \psi_I}{\partial x^2} = \frac{qN_A}{\varepsilon_{Si}} \qquad (14)$$

The boundary conditions can be written as:

$$\frac{\partial^2 \psi_{II}}{\partial x^2} = \frac{qN_D}{\varepsilon_{Si}} \qquad (15)$$

The boundary conditions can be written as:

$$\begin{aligned} \psi_I(-x_p) &= \psi_P \\ \psi_{II}(x_N) &= \psi_N \\ \psi_N - \psi_p &= V_{bi} \end{aligned} \qquad (16)$$

Outside the space region the charge density is zero so that

$$\frac{\partial^2 \psi}{\partial x^2} = 0 \qquad (17)$$

This implies that electric field outside the depletion region is constant. However, to be consistent, this electric field must be zero, otherwise it would imply a non-zero current, some applied bias etc.

$$\left.\frac{\partial \psi_I}{\partial x}\right|_{x=x_p} = 0$$

$$\left.\frac{\partial \psi_{II}}{\partial x}\right|_{x=x_N} = 0 \qquad (18)$$

- The electric field at x=0 must be continuous, otherwise it would imply an infinite charge density.

$$\left.\frac{\partial \psi_I}{\partial x} = \frac{\partial \psi_{II}}{\partial x}\right|_{x=0} \qquad (19)$$

- Similarly, the potential at x=0 also must be continuous

$$\psi_I(0) = \psi_{II}(0) \qquad (20)$$

- The Poisson equation with these boundary conditions can be easily solved to obtain the following results.

  Electric field: It is max. at the junction

$$\varepsilon_I(x) = \varepsilon_{\max} \frac{qN_A}{\varepsilon_{Si}} x \qquad (21)$$

$$\varepsilon_{II}(x) = \varepsilon_{\max} \frac{qN_D}{\varepsilon_{Si}} x \qquad (22)$$

$$\varepsilon_{max} = \sqrt{\frac{2q}{\varepsilon_{Si}}(\frac{N_D N_A}{N_D + N_A})V_{bi}} \quad \textit{at the junction} \quad (23)$$

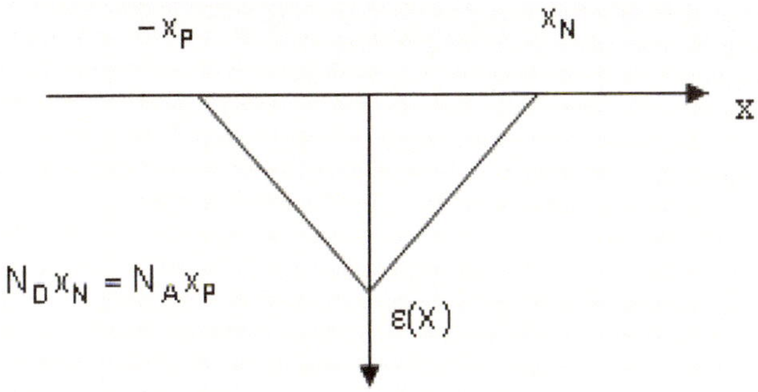


Potential:

$$\psi_I(x) = \psi_p + \frac{qN_A}{2\varepsilon_{Si}}(x + x_p)^2 \quad \textit{for} \ -x_p \le x \le 0 \quad (24)$$

$$\psi_{II}(x) = \psi_N + \frac{qN_D}{2\varepsilon_{Si}}(x \quad x_N)^2 \quad \textit{for} \ 0 \le x \le x_N \quad (25)$$

- The variation of potential across the depletion region is parabolic.

  Using the boundary condition that potential must be continuous at the junction

$$V_{bi} = \psi_N \quad \psi_p = \frac{q}{2\varepsilon_{Si}}(N_A x_p^2 + N_D x_N^2) \quad (26)$$

- Deletion widths: Using the relation $N_D x_N = N_A x_p$, we can obtain

$$W = x_p + x_N = \sqrt{\frac{2\varepsilon_{Si}}{q}(\frac{1}{N_D} + \frac{1}{N_A})V_{bi}} \quad (27)$$

$$x_p = \frac{W}{1 + \frac{N_A}{N_D}} \quad (28)$$

$$x_N = \frac{W}{1 + \frac{N_D}{N_A}} \quad (29)$$

- The depletion widths vary inversely with the doping.

## Comparison With Exact Numerical calculations

The Figure below shows a comparison of an actual charge profile computed using a 1-D device simulator and charge profile under depletion approximation for a doping of $(10^{17} cm^{-3})$.

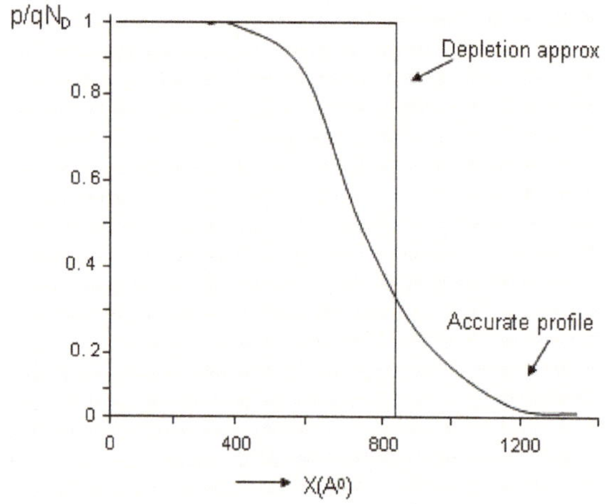


The Figure above shows that the transition region is about 600 $A^0$, almost same as the depletion width(735 $A^0$) predicted by the depletion approximation!

- The depletion approximation therefore appears to be a poor assumption. However, a careful look shows that the depletion assumption overestimates the charge in region I but underestimates the charge in region II. Since, the electric field and potential are determined by the integral of charge density, the error in electric field and potential profile is not large!

## Shockley Diode Equation

The *Shockley diode equation* or the *diode law*, named after transistor co-inventor William Shockley of Bell Telephone Laboratories, gives the I–V (current-voltage) characteristic of an idealized diode in either forward or reverse bias (applied voltage):

$$I = I_S \left( e^{\frac{V_D}{nV_T}} - 1 \right)$$

where

$I$ is the diode current,

$I_S$ is the reverse bias saturation current (or scale current),

$V_D$ is the voltage across the diode,

$V_T$ is the thermal voltage $kT/q$ (Boltzmann constant times temperature divided by electron charge), and

$n$ is the *ideality factor*, also known as the *quality factor* or sometimes *emission coefficient*.

The equation is called the *Shockley ideal diode equation* when $n$, the ideality factor, is set equal to 1. The ideality factor $n$ typically varies from 1 to 2 (though can in some cases be higher), depending on the fabrication process and semiconductor material and is set equal to 1 for the case of an "ideal" diode (thus the n is sometimes omitted). The ideality factor was added to account for imperfect junctions as observed in real transistors. The factor mainly accounts for carrier recombination as the charge carriers cross the depletion region.

The thermal voltage $V_T$ is approximately 25.85 mV at 300 K, a temperature close to "room temperature" commonly used in device simulation software. At any temperature it is a known constant defined by:

$$V_T = \frac{kT}{q},$$

where $k$ is the Boltzmann constant, $T$ is the absolute temperature of the p–n junction, and $q$ is the magnitude of charge of an electron (the elementary charge).

The reverse saturation current, $I_S$, is not constant for a given device, but varies with temperature; usually more significantly than $V_T$, so that $V_D$ typically decreases as $T$ increases.

The Shockley diode equation doesn't describe the "leveling off" of the I–V curve at high forward bias due to internal resistance. This can be taken into account by adding a resistance in series.

Under *reverse bias* (when the n side is put at a more positive voltage than the p side) the exponential term in the diode equation is near zero and the current is near a constant (negative) reverse current value of $-I_S$. The reverse *breakdown region* is not modeled by the Shockley diode equation.

For even rather small *forward bias* voltages the exponential is very large, since the thermal voltage is very small in comparison. The subtracted '1' in the diode equation is then negligible and the forward diode current can be approximated by

$$I = I_S e^{\frac{V_D}{nV_T}}$$

The use of the diode equation in circuit problems is illustrated in the article on diode modeling.

## Derivation

Shockley derives an equation for the voltage across a p-n junction in a long article published in 1949. Later he gives a corresponding equation for current as a function of voltage under additional assumptions, which is the equation we call the Shockley ideal diode equation. He calls it "a theoretical rectification formula giving the maximum rectification", with a footnote referencing a paper by Carl Wagner, *Physikalische Zeitschrift*.

To derive his equation for the voltage, Shockley argues that the total voltage drop can be divided into three parts:

- the drop of the quasi-Fermi level of holes from the level of the applied voltage at the p terminal to its value at the point where doping is neutral (which we may call the junction)
- the difference between the quasi-Fermi level of the holes at the junction and that of the electrons at the junction
- the drop of the quasi-Fermi level of the electrons from the junction to the n terminal.

He shows that the first and the third of these can be expressed as a resistance times the current, $R_1 I$. As for the second, the difference between the quasi-Fermi levels at the junction, he says that we can estimate the current flowing through the diode from this difference. He points out that the current at the p terminal is all holes, whereas at the n terminal it is all electrons, and the sum of these two is the constant total current. So the total current is equal to the decrease in hole current from one side of the diode to the other. This decrease is due to an excess of recombination of electron-hole pairs over generation of electron-hole pairs. The rate of recombination is equal to the rate of generation when at equilibrium, that is, when the two quasi-Fermi levels are equal. But when the quasi-Fermi levels are not equal, then the recombination rate is $\exp((\phi_p - \phi_n)/V_T)$ times the rate of generation. We then assume that most of the excess recombination (or decrease in hole current) takes place in a layer going by one hole diffusion length ($L_p$) into the n material and one electron diffusion length ($L_n$) into the p material, and that the difference between the quasi-Fermi levels is constant in this layer at $V_J$. Then we find that the total current, or the drop in hole current, is

$$I = I_s[e^{V_J/V_T} - 1]$$

where

$$I_s = gq(L_p + L_n)$$

and $g$ is the generation rate. We can solve for $V_J$ in terms of $I$:

$$V_J = V_T \ln\left(1 + \frac{I}{I_s}\right)$$

and the total voltage drop is then

$$V = IR_1 + V_T \ln\left(1 + \frac{I}{I_s}\right).$$

When we assume that $R_1$ is small, we obtain $V = V_J$ and the Shockley ideal diode equation.

The small current that flows under high reverse bias is then the result of thermal generation of electron-hole pairs in the layer. The electrons then flow to the n terminal and the holes to the p terminal. The concentrations of electrons and holes in the layer is so small that recombination there is negligible.

In 1950 Shockley and coworkers published a short article describing a germanium diode that closely followed the ideal equation.

In 1954, Bill Pfann and W. van Roosbroek (who were also of Bell Telephone Laboratories) reported that while Shockley's equation was applicable to certain germanium junctions, for many silicon junctions the current (under appreciable forward bias) was proportional to $e^{V_J/(AV_T)}$, with $A$ having a value as high as 2 or 3. This is the "ideality factor" called $n$ above.

In 1981, Alexis de Vos and Herman Pauwels showed that a more careful analysis of the quantum mechanics of a junction, under certain assumptions, gives a current versus voltage characteristic of the form

$$I(V) = -qA[F_i - 2F_o(V)]$$

in which $A$ is the cross-sectional area of the junction and $F_i$ is the number of in-coming photons per unit area, per unit time, with energy over the band-gap energy, and $F_o(V)$ is out-going photons, given by

$$F_o(V) = \int_{v_g}^{\infty} \frac{1}{\exp\left(\frac{hv - qV}{kT_c}\right) - 1} \frac{2\pi v^2}{c^2} dv.$$

Although this analysis was done for photovoltaic cells under illumination, it applies also when the illumination is simply background thermal radiation. It gives a more rigorous form of expression for ideal diodes in general, except that it assumes that the cell is thick enough that it can produce this flux of photons. When the illumination is just background thermal radiation, the characteristic is

$$I(V) = 2q[F_o(V) - F_o(0)]$$

Note that, in contrast to the Shockley law, the current goes to infinity as the voltage goes to the gap voltage $hv_g/q$. This of course would require an infinite thickness to provide an infinite amount of recombination.

## I - V Characteristics in Reverse Bias

- A PN Junction is said to be in Forward Bias when the P-type region (Anode) is made positive with respect to the N-type region (Cathode).
- A PN Junction is said to be in Reverse Bias when the P-type region (Anode) is made negative with respect to the N-type region (Cathode).

Let us consider the Forward bias first and examine qualitatively the mode of operation

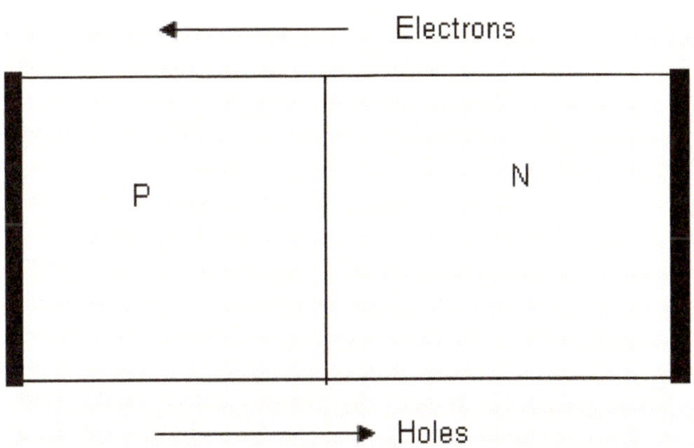


The holes are required to move from $P \rightarrow N$ and electrons from $N \rightarrow P$ .

- There are plenty of holes in P-type region and would like to move to N-region via diffusion but are prevented by the electric field (or the energy barrier) at equilibrium. The drift and diffusion currents cancel each other

$$I_P(diff.) - I_P(drift) = 0 \quad (30)$$

- Similarly, there are plenty of electrons in N-type region and would like to move to P-region via diffusion but are prevented by the electric field (or the energy barrier) at equilibrium. The drift and diffusion currents again cancel each other.

$$I_N(diff.) - I_N(drift) = 0 \quad (31)$$

- The application of forward bias reduces the barrier and the electric field allowing significant electron and hole current to flow:

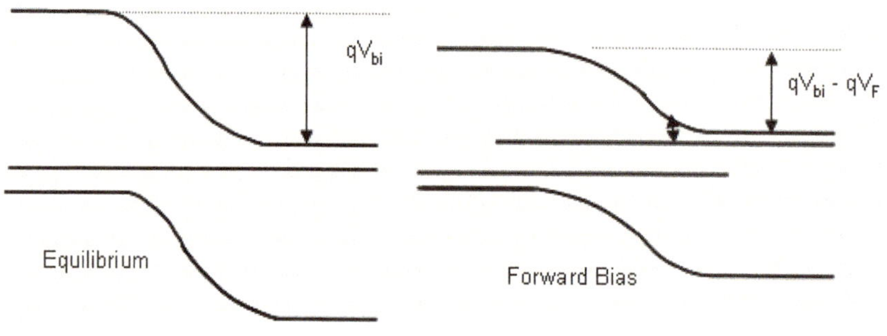


- The fraction of electrons that are able to cross over to the P-side or the fraction of holes that are able to cross over to the N-side and contribute to current goes exponentially with the barrier height (remember, $n \propto e^{\frac{q\psi}{kT}}$)
$\Rightarrow$ Current increases exponentially with the applied forward bias.

## Reverse Bias

- The holes are now required by the applied bias to move from $N \rightarrow P$ and electrons from $P \rightarrow N$ as shown below:

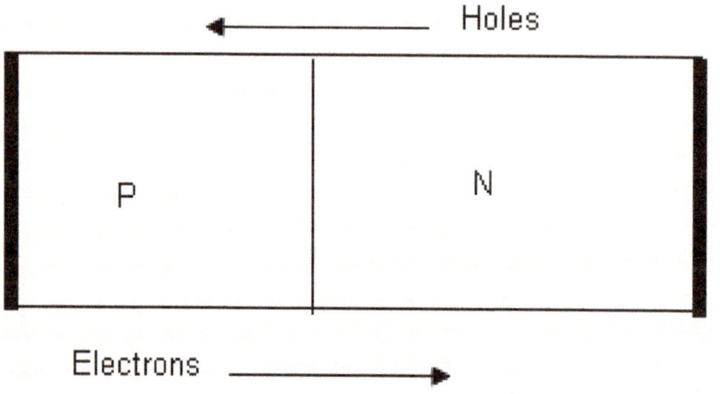


- Although the electric field favors the flow of holes to the P-region, there are very few holes in N-region to begin with! The number of holes in N-region is $\frac{n_i^2}{N_D}$, a very small number.

- Further, the number of holes is fixed and unaffected by the bias.
- Similarly, the number of available electrons in P-region for current flow is very small and unaffected by the applied bias.
- The only thing that the applied reverse bias does is to increase the junction electric field or the barrier height as shown below

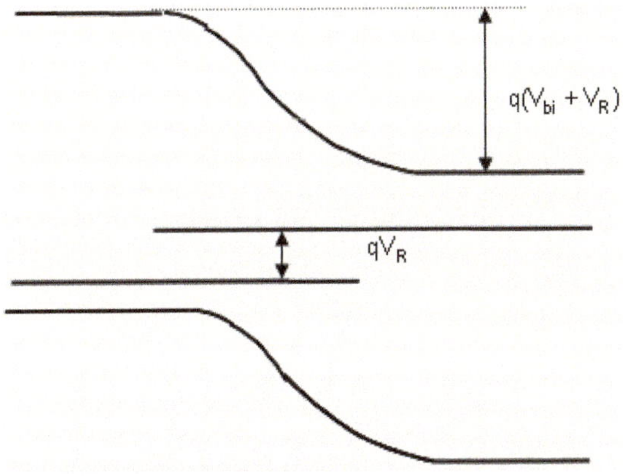


The increased electric field does not alter the current flow because the bottleneck is the small number of carriers available for current conduction.

⇒Current in Reverse bias is very small and almost constant

## Static I-V Characteristics

The dc current-voltage characteristics of the PN junction diode will be obtained using the semiconductor equations listed below:

$$J_N = qn\mu_N\varepsilon + qD_N\frac{\partial n}{\partial x} \quad ; \quad J_P = qp\mu_P\varepsilon + qD_p\frac{\partial p}{\partial x} \qquad (32)$$

$$\frac{\partial^2\psi}{\partial x^2} = \frac{q}{\varepsilon_{Si}}(N_D + p(x) \quad n(x) \quad N_A) \qquad (33)$$

$$\frac{\partial n}{\partial t} = \frac{1}{q}\frac{\partial J_N}{\partial x} + G_N \quad R_N \quad ; \quad \frac{\partial p}{\partial t} = \frac{1}{q}\frac{\partial J_p}{\partial x} + G_p \quad R_p \qquad (34)$$

$$n = n_i\exp(\frac{E_{FN} \quad E_i}{kT}) \quad ; \quad p = n_i\exp(\frac{E_i \quad E_{FP}}{kT}) \qquad (35)$$

$$np = n_i^2\exp(\frac{E_{FN} \quad E_{FP}}{kT}) \qquad (36)$$

$$J_N = n\mu_N\frac{\partial E_{FN}}{\partial x} \quad ; \quad J_p = n\mu_p\frac{\partial E_{FP}}{\partial x} \qquad (37)$$

In steady state, the continuity equation reduces to

$$O = \frac{1}{q}\frac{\partial J_N}{\partial x} + G_N - R_N \quad ; O = \frac{1}{q}\frac{\partial J_P}{\partial x} + G_P - R_P \qquad (38)$$

- Since for every electron lost/generated due to recombination/generation, there is a corresponding hole lost/generated also

$$\Rightarrow G_N = G_p \quad ; \quad R_N = R_P \qquad (39)$$

$$\Rightarrow \frac{1}{q}\frac{\partial(J_N + J_P)}{\partial x} = 0 \qquad (40)$$

$$\Rightarrow J_{Total} = J_N(x) + J_P(x) = cons\tan t \qquad (41)$$

In other words, the net current flowing through the device is the same everywhere.

- Since the current is the same everywhere, one can choose the region within the device for calculation of current-voltage characteristics.

  Big Question : Where in the device should the current be calculated such that its computation besides being easy is also accurate ?

- Let us consider some alternatives: (i) At the junction:
- To appreciate the ease or difficulty of carrying out the computation in this case, let us consider a symmetric junction with

$$N_D = N_A = 10^{17} cm^3$$

The net electron current is equal to: $J_N = qn\mu_N\varepsilon + qD_N\frac{\partial n}{\partial x}$

The drift and diffusion currents oppose each other so that

$$J_N = J_N(diff.) - J_N(drift) \qquad (42)$$

Let us try to estimate the magnitude of the drift component:

- Because of symmetry n(o) = p(o) at the junction

Further : $np = n_i^2 \exp(\frac{E_{FN} - E_{FP}}{kT})$

$$\Rightarrow n(0) = n_i \exp(\frac{E_{FN} - E_{FP}}{kT}) \qquad (43)$$

Assumption (i) : All the voltage is dropped across the junction:

$\Rightarrow E_{FN} - E_{FP} = qV$ with the junction

Net voltage across the junction $= V_{bi} - V$

Assumption (ii) : Depletion approximation

$$\Rightarrow \varepsilon_{\max} = \sqrt{\frac{q}{\varepsilon_{Si}}\frac{1}{N_D}(V_{bi} - V)} \qquad (44)$$

- For a forward bias of o.6V, the electron drift current can be calculated using the results obtained as equal to $\cong 10^3 A/cm^2$

- As we shall see later, the net electron current flowing through the junction for this device at a forward bias of 0.6V is $\cong 10^{-2} A/cm^2$

$$\Rightarrow J_N(diff.) - J_N(drift) = 10^{-2} \qquad (45)$$

Because the drift current ($\cong 10^3 A/\text{cm}^2$) is five orders of magnitude larger than the net current, the drift and diffusion currents would have to be calculated to an accuracy of .001% to obtain a correct estimate of the net electron current!

⇒ This makes the estimation of total current via an analysis at the junction virtually impossible!

- Let us consider a region for estimation of current which is far from the junction in say N-type semiconductor.
- Far from the junction, on the N-side, the current is expected to be primarily an electron current. Any holes which are injected from the P-side would recombine and disappear away from the junction.
- The electron density being constant, the electron current would be primarily a drift current so that

$$J_{Total} = qN_D\mu_N\varepsilon \qquad (46)$$

It might appear that this is a very good place for estimation of current because we have just one component and only one unknown , the electric field .

- However, this electric field is extremely difficult to estimate because of its very small value.

The voltage applied across the diode gets dropped partially across the junction and partially outside it

$$V = V_J + V_P + V_N \qquad (47)$$

where the last two terms represent the voltage dropped across the neutral N and P-regions

- The bottleneck for current flow in a PN junction is the space charge region where the potential barrier exists. As a result, is almost equal to the applied voltage

$$V - V_J = V_P + V_N \qquad (48)$$

While it is easy to compute the junction voltage fairly accurately, the estimation of residual drops in the neutral regions becomes very difficult.

- The two examples discussed earlier illustrate that the choice of position in the PN junction for computation of its I-V characteristics is very important.
- As first demonstrated by Shockley, the computation of currents in PN junction diode is best done at the edges of depletion region as explained below:

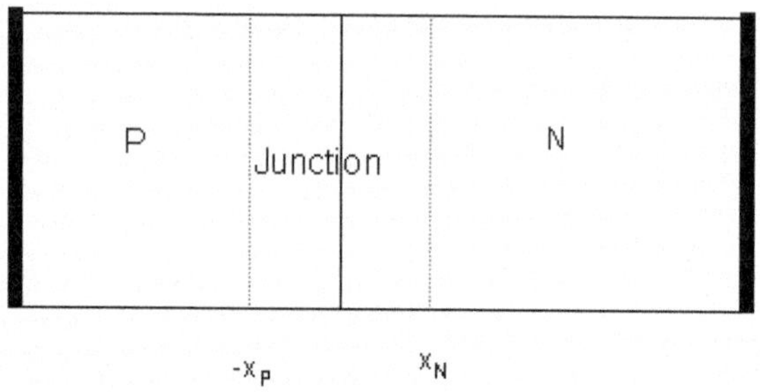


$$J_{Total}(x_N) = J_N(x_N) + J_P(x_N) \quad (49)$$
$$J_{Total}(-x_P) = J_N(-x_P) + J_P(-x_P) \quad (50)$$

- During the course of the analysis, several assumptions will be made. There are two ways of justifying these assumptions. One of them is:

  (i) Make the assumption

  (ii) Solve the resulting simplified equations to obtain the current-voltage characteristics

  (iii) Check that the assumptions made are consistent with the results obtained.

The assumptions made will be consistent only for certain range of currents, so that the range of validity of the model will be obtained.

- The other approach is to justify the assumptions in the beginning of the analysis, based on available device characteristics. These assumptions would define the range of validity of the obtained model.

We shall follow a mix of these two approaches

Assumption (1): Negligible recombination within the Junction

We shall justify this assumption using the first approach, namely that the assumption would be shown to be consistent with the results obtained within certain limits.

$\Rightarrow$ All the ho;es that are injected at $x = -x_P$ reach the point $x = x_N$ so that

$$J_P(-x_P) = J_P(x_N) \quad (51)$$

Similarly all the electrons that are injected at $x = x_N$ reach the point $x = -x_P$ , so that

$$J_N(x_N) = J_N(-x_P) \quad (52)$$

This allows the total current to be expressed as :

$$J_{Total} = J_N(-x_P) + J_P(x_N) \quad (53)$$

The total current can be computed by computing the minority carrier currents at the edges of depletion region in N and P-regions

Assumption (2) : Minority carrier current is largely diffusive

We shall justify this assumption using the second approach, namely that the validity of this assumption will be demonstrated prior to analysis. This is described in Appendix A.

The assumption implies :

$$J_P(x) \cong J_P(\text{diff.}) \text{ for } x \geq x_N \quad (54)$$

$$J_N(x) \cong J_N(\text{diff.}) \text{ for } x \leq -x_P \quad (55)$$

$$\Rightarrow J_{Total} = -qD_P \left.\frac{\partial p}{\partial x}\right|_{x=x_N} + qD_N \left.\frac{\partial n}{\partial x}\right|_{x=-x_P} \qquad (56)$$

⇒The task of computing the currents boils down to the computation of minority carrier profiles: p(x) in N-region and n(x) in P-region.

- The minority carrier profile can be determined by solving the continuity equation with appropriate boundary conditions

  For hole density in N-region:

$$0 = -\frac{1}{q}\frac{\partial J_P}{\partial x} + G_P - R_P \qquad (57)$$

- In Silicon, the dominant recombination mechanism is the Shockley-Hall-Read recombination which can be described by the relation

$$U = R_P - G_P \cong \frac{\Delta P}{\tau_P} \qquad (58)$$

under low level injection conditions. $\tau$ is the hole recombination lifetime in N-type material.

The hole continuity equation can be re-written as

$$\frac{\partial^2 p(x)}{\partial x^2} - \frac{\partial p(x)}{L_P^2} = 0 \qquad (59)$$

where $L_p = \sqrt{D_P T_P}$ has the units of length and as we shall see later is appropriately called the hole diffusion length

Boundary Conditions:

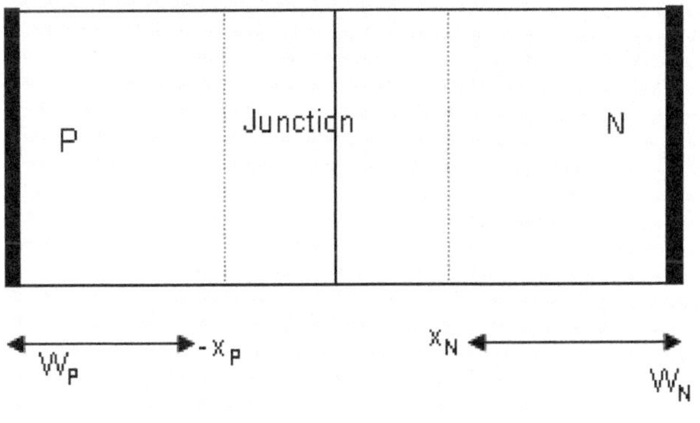


$$\Delta p(x_N) = \Delta p_0 \qquad (60)$$

$$\Delta p(x_N + W_N) = 0 \qquad \text{assuming ideal ohmic contact}$$

Solution:

$$\Delta p(x) = \Delta p_0 \frac{\sinh(\frac{W_N - (x - x_N)}{L_P})}{\sinh \frac{W_N}{L_P}} \quad \text{for } x \geq x_N \quad (61)$$

Similarly for the N-side:

$$\frac{\partial^2 n(x)}{\partial x^2} - \frac{\partial n(x)}{L_N^2} = 0 \qquad (62)$$

Where $L_N = \sqrt{D_N T_N}$ is called the electron diffusion length.

Boundary conditions:

$$\Delta n(-x_P) = \Delta n_0 \qquad (63)$$

$$\Delta n(-x_p + W_p) = 0 \qquad \text{assuming ideal ohmic contact}$$

$$\Delta n(x) = \Delta n_0 \frac{\sinh(\frac{W_P - (x + x_N)}{L_N})}{\sinh \frac{W_P}{L_N}} \quad \text{for } x \leq -x_P \quad (64)$$

There are two extreme cases:

(i) Wide Base diode: $W_N \gg L_P$ ; $W_P \gg L_N$

For this case, the minority carrier densities can be simplified to:

$$\Delta p(x) = \Delta p_0 \exp(-\frac{x - x_N}{L_P}) \quad (65)$$

$$\Delta n(x) = \Delta n_0 \exp(\frac{x + x_P}{L_P}) \quad (66)$$

⇒The minority carrier densities decay exponentially with the distance from the junction, with a characteristics decay length of $L_P$ for holes and $L_N$ for electrons.

- It can be shown that the average distance a hole diffuses before recombining is equal to $L_P$ so that it is called the diffusion length.

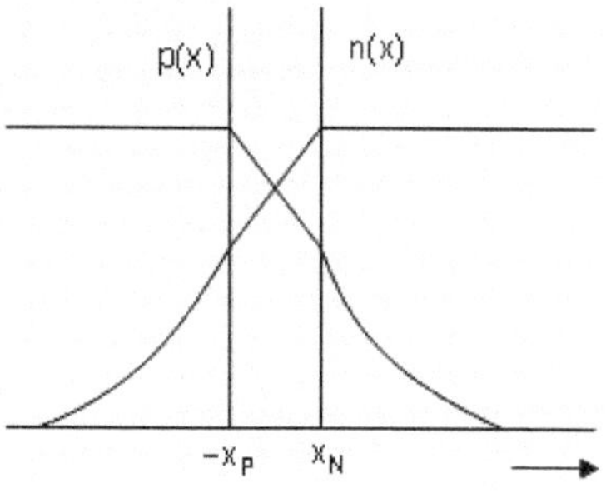

The other extreme case is : $W_N \ll L_P; W_P \ll L_N$

(ii) Narrow Base diode:

The minority carrier profile can be simplified to

$$\Delta p(x) = \Delta p_0 \frac{W_N - (x - x_N)}{W_N} \quad (67)$$

$$\Delta n(x) = \Delta n_0 \frac{W_P + (x + x_P)}{W_P} \quad (68)$$

The carrier densities vary linearly with position now !

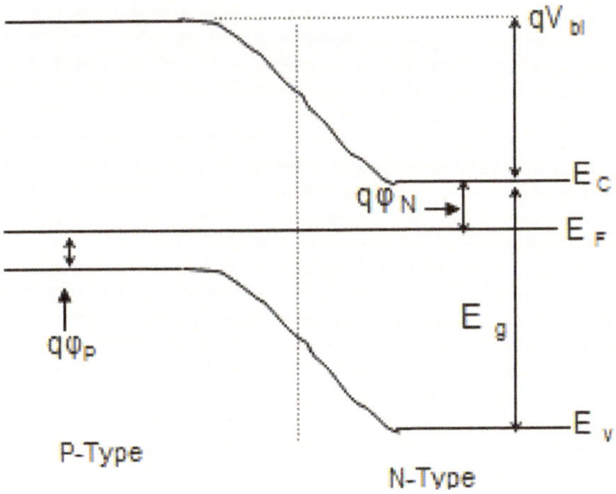


- The total diode current for wide and narrow base diodes can be expressed as

  Wide base diode:

$$J_{Total} = q\frac{D_P}{L_P}\Delta p_0 + q\frac{D_N}{L_N}\Delta n_0 \quad (69)$$

  Narrow base diode:

$$J_{Total} = q\frac{D_P}{W_N}\Delta p_0 + q\frac{D_N}{W_P}\Delta n_0 \quad (70)$$

$\Rightarrow$The task of determining the I-V Characteristics now reduces to finding a relationship between the minority carrier densities at the edges of depletion region and the applied voltage.

We start with the relation:

$$P(x_N)n(x_N) = n_i^2 \exp(\frac{E_{FN}(x_N) - E_{FP}(x_N)}{kT}) \quad (71)$$

where quasi-neutrality$\Rightarrow$ $n(x_N) = N_D + P(x_N)$ (72)

$$P(x_N)^2 + p(x_N)N_D = n_i^2 \exp(\frac{E_{FN}(x_N) - E_{FP}(x_N)}{kT}) \quad (73)$$

The low level injection assumption invoked earlier can be used here also for simplification. The first obvious consequence is that

$$P(x_N) \ll n(x_N) \Rightarrow P(x_N) \ll N_D \quad (74)$$

So that the first term on the LHS of the above expression can be neglected.

The second consequence of low level injection, explained in detail in Appendix A is that

- $E_{FN} \cong constant$ for $x \le x_N$ in the N-region and the depletion region
- $E_{FP} \cong constant$ for $x \ge -x_P$ in the P-region and the depletion region
- The quasi-Fermi level $E_{FN}$ on the N-side must coincide with the Fermi level of the metal forming the ohmic contact to the N-side if an ideal contact with no voltage drop across it is assumed.
- Similarly, the quasi-Fermi level $E_{FP}$ on the P-side must coincide with the Fermi level of the metal forming the ohmic contact to the P-side if an ideal contact with no voltage drop across it is assumed.

Since a voltage V is applied between the two ohmic contacts:

$$\Rightarrow E_{FN} - E_{FP} = qV \quad (75)$$

This allows the minority carrier densities at the edges of depletion region to be expressed as

$$\Delta p_0 = \frac{n_i^2}{N_D}(\exp(\frac{qV}{kT}) - 1) \quad (76)$$

$$\Delta n_0 = \frac{n_i^2}{N_A}(\exp(\frac{qV}{kT}) - 1) \quad (77)$$

- The total current density for the diode at a bias of V volts can now be expressed as

  Wide base diode:

$$J_{Total} = qn_i^2(\frac{D_P}{L_P N_D} + \frac{D_N}{L_N N_A})(\exp(\frac{qV}{kT}) - 1) \quad (78)$$

  Narrow base diode:

$$J_{Total} = qn_i^2(\frac{D_P}{W_N N_D} + \frac{D_N}{W_P N_A})(\exp(\frac{qV}{kT}) - 1) \quad (79)$$

- The current varies exponentially with applied voltage when the diode is forward biased (V > o)
- The current is constant and small when the diode is reverse biased (V < o)

## I - V Characteristics in Forward Bias

- The equation for current flowing through the diode, derived earlier, is given by the expression:

$$J_{Total} = qn_i^2\left(\frac{D_N}{L_N N_A} + \frac{D_P}{L_P N_D}\right)\left(\exp\left(\frac{qV}{kT}\right) - 1\right) \quad (1)$$

This expression was derived under the assumptions:

(i) Low Level Injection

(ii) Negligible recombination within the SCR

Although the equation was derived in the context of forward bias, much of the derivation remains valid in reverse bias also

- In reverse bias, instead of injection of minority carriers in P and N-regions, there is extraction of minority carriers from them.

Holes now flow from $N \rightarrow P$ and electrons from $P \rightarrow N$ .

- As a result, N-region gets depleted of holes and P-region gets depleted of electrons

$$\Delta p(x) = p(x) - \frac{n_i^2}{N_D} < 0 \quad \text{in the N} - \text{region} \quad (2)$$

$$\Delta n(x) = n(x) - \frac{n_i^2}{N_A} < 0 \quad \text{in the P} - \text{region} \quad (3)$$

Since $\Delta p(x)$ and $\Delta n(x)$ can have a maximum value of $-\frac{n_i^2}{N_D}$ *and* $-\frac{n_i^2}{N_A}$ and respectively,

$$\Rightarrow \Delta p(x) << N_D$$
$$\Delta n(x) << N_A \quad (4)$$

These are the conditions for low level injection if "injection" is interpreted as having a negative value in this case

$$\Rightarrow J_P(X) \cong J_P(diff.)$$

$$= -qD_P \frac{\partial p}{\partial x} \quad (5)$$

$$\Rightarrow \Delta p(x) = \Delta p_0 \exp(-\frac{x - x_N}{L_P}) \quad (6)$$

$$\Delta p_0 = \frac{n_i^2}{N_D}(\exp(\frac{qV}{kT}) - 1) \quad (7)$$

$$= -\frac{n_i^2}{N_D} \quad \textit{because V is negative}$$

$$J_P(x_N) = -qn_i^2 \frac{D_P}{L_P N_D} \quad (8)$$

Similarly, for the N-side

$$\Rightarrow J_N(x) \cong J_N(\text{diff.})$$

$$= qD_N \frac{\partial n}{\partial x} \quad (9)$$

$$\Rightarrow \Delta n(x) = \Delta n_0 \exp\left(\frac{x + x_P}{L_N}\right) \quad (10)$$

$$\Delta \quad = \frac{}{N}(\exp(\frac{qV}{kT}) - 1) \quad (11)$$

$$= -\frac{n_i^2}{N_A} \quad \textit{because V is negative}$$

$$J_N(-x_P) = -qn_i^2 \frac{D_N}{L_N N_A} \quad (12)$$

The spatial variation of electron and hole density is shown below:

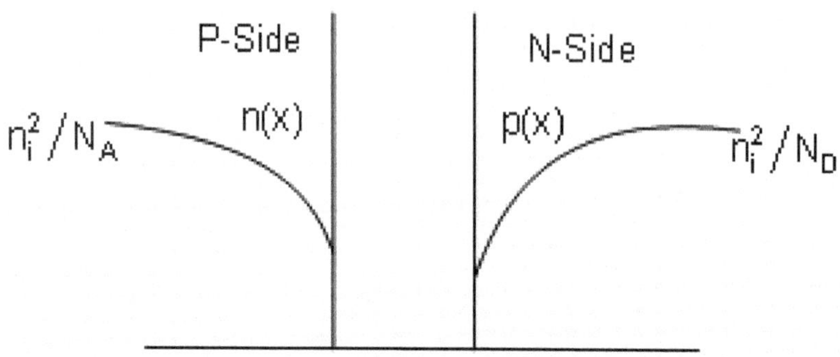


- The electrons diffuse from the bulk of the P-region to the edge of the depletion region after which they are swept away by the junction field. Similarly, the holes diffuse from the bulk of the N-region to the depletion region edge after which they are swept by the electric field to the P-region.

- The source of electrons in P-region and holes in N-region is thermal generation of carriers.

It was shown earlier, in the context of forward bias, that :

$$J_{Total} = J_P(x_N) + J_N(-x_P) + J_R \qquad (13)$$

- This expression is equally valid in reverse bias also, with the difference that the last term now represents generation of carriers within the space charge region, instead of recombination.
- In Forward bias, we had neglected this term but as we shall see, this term is the dominant term under reverse bias for Silicon PN junction diodes.

$$J_R = q\int_{-x_P}^{x_N} U dx \qquad (14)$$

$$U = \frac{pn - n_i^2}{p + n + 2n_i}\tau^{-1} \qquad (15)$$

Since $E_{FN} - E_{FP} < 0$ within the space charge region:

$$pn = n_i^2 \exp(\frac{E_{FN} - E_{FP}}{kT}) \ll n_i^2 \qquad (16)$$

$$G = -U = \frac{n_i^2}{p + n + 2n_i}\tau^{-1} \qquad (17)$$

The negative sign indicates generation !

- Over a large fraction of the depletion width, the electron and hole densities are much smaller than the intrinsic carrier density so that

Approximation: $G \cong -\frac{n_i}{2\tau}$ for $-x_P \leq x \leq x_N$ (18)

This allows the generation current to be written as :

$$J_R = -q\frac{n_i}{2\tau}W \qquad (19)$$

The net reverse bias current can be written as:

$$J_{Total} = -qn_i^2(\frac{D_P}{L_P N_D} + \frac{D_N}{L_N N_A}) - q\frac{n_i}{2\tau}W \qquad (20)$$

The first tem represents the current due to minority carrier diffusion and the second due to generation within the space charge region.

# Permissions

All chapters in this book are published with permission under the Creative Commons Attribution Share Alike License or equivalent. Every chapter published in this book has been scrutinized by our experts. Their significance has been extensively debated. The topics covered herein carry significant information for a comprehensive understanding. They may even be implemented as practical applications or may be referred to as a beginning point for further studies.

We would like to thank the editorial team for lending their expertise to make the book truly unique. They have played a crucial role in the development of this book. Without their invaluable contributions this book wouldn't have been possible. They have made vital efforts to compile up to date information on the varied aspects of this subject to make this book a valuable addition to the collection of many professionals and students.

This book was conceptualized with the vision of imparting up-to-date and integrated information in this field. To ensure the same, a matchless editorial board was set up. Every individual on the board went through rigorous rounds of assessment to prove their worth. After which they invested a large part of their time researching and compiling the most relevant data for our readers.

The editorial board has been involved in producing this book since its inception. They have spent rigorous hours researching and exploring the diverse topics which have resulted in the successful publishing of this book. They have passed on their knowledge of decades through this book. To expedite this challenging task, the publisher supported the team at every step. A small team of assistant editors was also appointed to further simplify the editing procedure and attain best results for the readers.

Apart from the editorial board, the designing team has also invested a significant amount of their time in understanding the subject and creating the most relevant covers. They scrutinized every image to scout for the most suitable representation of the subject and create an appropriate cover for the book.

The publishing team has been an ardent support to the editorial, designing and production team. Their endless efforts to recruit the best for this project, has resulted in the accomplishment of this book. They are a veteran in the field of academics and their pool of knowledge is as vast as their experience in printing. Their expertise and guidance has proved useful at every step. Their uncompromising quality standards have made this book an exceptional effort. Their encouragement from time to time has been an inspiration for everyone.

The publisher and the editorial board hope that this book will prove to be a valuable piece of knowledge for students, practitioners and scholars across the globe.

# Index